Fundamentals of Thermodynamics

Fundamentals of Thermodynamics

Edited by
Henry Medina

Larsen & Keller
www.larsen-keller.com

Fundamentals of Thermodynamics
Edited by Henry Medina
ISBN: 978-1-63549-688-8 (Hardback)

Larsen & Keller

Published by Larsen and Keller Education,
5 Penn Plaza,
19th Floor,
New York, NY 10001, USA

Cataloging-in-Publication Data

Fundamentals of thermodynamics / edited by Henry Medina.
p. cm.
Includes bibliographical references and index.
ISBN 978-1-63549-688-8
1. Thermodynamics. 2. Heat. 3. Heat-engines. I. Medina, Henry.
TJ265 .F86 2018
621.402 1--dc23

Printed and bound in China.

For more information regarding Larsen and Keller Education and its products, please visit the publisher's website www.larsen-keller.com

Table of Contents

Preface

Thermodynamics, as a sub-branch of physics, refers to the study of the interrelation between energy, work, heat and temperature. It is based on the four major laws of thermodynamics and is divided into four major parts, namely, chemical thermodynamics, classical thermodynamics, equilibrium treatment and statistical mechanism. The topics covered in this text offer the readers new insights in the field of thermodynamics. Different approaches, evaluations and methodologies have been included in it. This textbook is an essential guide for both academicians and those who wish to pursue this discipline further.

A short introduction to every chapter is written below to provide an overview of the content of the book:

Chapter 1 - Thermodynamics is the study of the relation between work, heat, and properties of a material. This relation is directed by the four laws of thermodynamics. The concept has its application in various devices such as automotive engines, refrigerators, air-conditioners, etc. The chapter on thermodynamics offers an insightful focus, keeping in mind the complex subject matter; **Chapter 2** - Work is considered as a mode of energy transfer. Work takes place at the boundary of the system and can be observed whenever a system is undergoing a process. Another concept explored in this chapter is heat, which is the energy that flows from one body to another due to an occurrence of change in temperatures. Work and heat are one of the most significant and important topics related to thermodynamics. Different types of work include shaft work, spring work, work on elastic solids, etc. This chapter elucidates the crucial theories and principles of thermodynamics; **Chapter 3** - The first law of thermodynamics follows the law of conservation of energy, which states that energy can neither be created nor destroyed. As per the first law, the increase in energy is equal to heat let out by the system. The second law states that the heat will flow from the hotter body to the colder body until the temperatures become equal. The major components of the first and second laws of thermodynamics are discussed in this section; **Chapter 4** - Energy is the capacity to do work. Energy, on the macroscopic level, can be divided into two kinds, kinetic and potential. The bifurcation of energy on the microscopic level includes translational energy, rotational energy, electronic energy, nuclear energy, etc. The chapter serves as a source to understand the major categories related to energy conversion processes; **Chapter 5** - Thermodynamics studies the conversion of energy from one form to another. The laws of thermodynamics are applied to almost every stream in science and technology. The topic can be classified into classical thermodynamics, statistical thermodynamics, kinetic theory and irreversible thermodynamics. The aspects elucidated in this chapter are of vital importance, and provide a better understanding of thermodynamics.

I extend my sincere thanks to the publisher for considering me worthy of this task. Finally, I thank my family for being a source of support and help.

Editor

1

An Overview of Thermodynamics

Thermodynamics is the study of the relation between work, heat, and properties of a material. This relation is directed by the four laws of thermodynamics. The concept has its application in various devices such as automotive engines, refrigerators, air-conditioners, etc. The chapter on thermodynamics offers an insightful focus, keeping in mind the complex subject matter.

Thermodynamics

Thermodynamics is a branch of physics concerned with heat and temperature and their relation to energy and work. The behavior of these quantities is governed by the four laws of thermodynamics, irrespective of the composition or specific properties of the material or system in question. The laws of thermodynamics are explained in terms of microscopic constituents by statistical mechanics. Thermodynamics applies to a wide variety of topics in science and engineering, especially physical chemistry, chemical engineering and mechanical engineering.

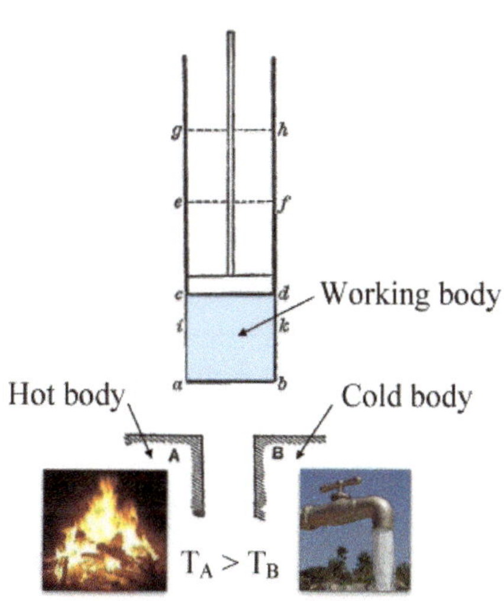

Annotated color version of the original 1824 Carnot heat engine showing the hot body (boiler), working body (system, steam), and cold body (water), the letters labeled according to the stopping points in Carnot cycle.

Historically, thermodynamics developed out of a desire to increase the efficiency of early steam engines, particularly through the work of French physicist Nicolas Léonard Sadi

Carnot (1824) who believed that engine efficiency was the key that could help France win the Napoleonic Wars. Scottish physicist Lord Kelvin was the first to formulate a concise definition of thermodynamics in 1854 which stated, *"Thermo-dynamics is the subject of the relation of heat to forces acting between contiguous parts of bodies, and the relation of heat to electrical agency."*

The initial application of thermodynamics to mechanical heat engines was extended early on to the study of chemical compounds and chemical reactions. Chemical thermodynamics studies the nature of the role of entropy in the process of chemical reactions and has provided the bulk of expansion and knowledge of the field. Other formulations of thermodynamics emerged in the following decades. Statistical thermodynamics, or statistical mechanics, concerned itself with statistical predictions of the collective motion of particles from their microscopic behavior. In 1909, Constantin Carathéodory presented a purely mathematical approach to the field in his axiomatic formulation of thermodynamics, a description often referred to as *geometrical thermodynamics.*

Introduction

A description of any thermodynamic system employs the four laws of thermodynamics that form an axiomatic basis. The first law specifies that energy can be exchanged between physical systems as heat and work. The second law defines the existence of a quantity called entropy, that describes the direction, thermodynamically, that a system can evolve and quantifies the state of order of a system and that can be used to quantify the useful work that can be extracted from the system.

In thermodynamics, interactions between large ensembles of objects are studied and categorized. Central to this are the concepts of the thermodynamic *system* and its *surroundings*. A system is composed of particles, whose average motions define its properties, and those properties are in turn are related to one another through equations of state. Properties can be combined to express internal energy and thermodynamic potentials, which are useful for determining conditions for equilibrium and spontaneous processes.

With these tools, thermodynamics can be used to describe how systems respond to changes in their environment. This can be applied to a wide variety of topics in science and engineering, such as engines, phase transitions, chemical reactions, transport phenomena, and even black holes. The results of thermodynamics are essential for other fields of physics and for chemistry, chemical engineering, aerospace engineering, mechanical engineering, cell biology, biomedical engineering, materials science, and economics, to name a few.

This article is focused mainly on classical thermodynamics which primarily studies systems in thermodynamic equilibrium. Non-equilibrium thermodynamics is often treated as an extension of the classical treatment, but statistical mechanics has brought many advances to that field.

The thermodynamicists representative of the original eight founding schools of thermodynamics. The schools with the most-lasting effect in founding the modern versions of thermodynamics are the Berlin school, particularly as established in Rudolf Clausius's 1865 textbook *The Mechanical Theory of Heat*, the Vienna school, with the statistical mechanics of Ludwig Boltzmann, and the Gibbsian school at Yale University, American engineer Willard Gibbs' 1876 *On the Equilibrium of Heterogeneous Substances* launching chemical thermodynamics.

History

The history of thermodynamics as a scientific discipline generally begins with Otto von Guericke who, in 1650, built and designed the world's first vacuum pump and demonstrated a vacuum using his Magdeburg hemispheres. Guericke was driven to make a vacuum in order to disprove Aristotle's long-held supposition that 'nature abhors a vacuum'. Shortly after Guericke, the English physicist and chemist Robert Boyle had learned of Guericke's designs and, in 1656, in coordination with English scientist Robert Hooke, built an air pump. Using this pump, Boyle and Hooke noticed a correlation between pressure, temperature, and volume. In time, Boyle's Law was formulated, which states that pressure and volume are inversely proportional. Then, in 1679, based on these concepts, an associate of Boyle's named Denis Papin built a steam digester, which was a closed vessel with a tightly fitting lid that confined steam until a high pressure was generated.

Later designs implemented a steam release valve that kept the machine from exploding. By watching the valve rhythmically move up and down, Papin conceived of the idea of a piston and a cylinder engine. He did not, however, follow through with his design. Nevertheless, in 1697, based on Papin's designs, engineer Thomas Savery built the first engine, followed by Thomas Newcomen in 1712. Although these early engines were crude and inefficient, they attracted the attention of the leading scientists of the time.

The fundamental concepts of heat capacity and latent heat, which were necessary for the development of thermodynamics, were developed by Professor Joseph Black at the University of Glasgow, where James Watt was employed as an instrument maker. Black and Watt performed experiments together, but it was Watt who conceived the idea of the external condenser which resulted in a large increase in steam engine efficiency. Drawing on all the previous work led Sadi Carnot, the "father of thermodynamics", to publish *Reflections on the Motive Power of Fire* (1824), a discourse on heat, power, energy and engine efficiency. The paper outlined the basic energetic relations between the Carnot engine, the Carnot cycle, and motive power. It marked the start of thermodynamics as a modern science.

The first thermodynamic textbook was written in 1859 by William Rankine, originally trained as a physicist and a civil and mechanical engineering professor at the University of Glasgow. The first and second laws of thermodynamics emerged simultaneously in the 1850s, primarily out of the works of William Rankine, Rudolf Clausius, and William Thomson (Lord Kelvin).

The foundations of statistical thermodynamics were set out by physicists such as James Clerk Maxwell, Ludwig Boltzmann, Max Planck, Rudolf Clausius and J. Willard Gibbs.

During the years 1873-76 the American mathematical physicist Josiah Willard Gibbs published a series of three papers, the most famous being *On the Equilibrium of Heterogeneous Substances*, in which he showed how thermodynamic processes, including chemical reactions, could be graphically analyzed, by studying the energy, entropy, volume, temperature and pressure of the thermodynamic system in such a manner, one can determine if a process would occur spontaneously. Also Pierre Duhem in the 19th century wrote about chemical thermodynamics. During the early 20th century, chemists such as Gilbert N. Lewis, Merle Randall, and E. A. Guggenheim applied the mathematical methods of Gibbs to the analysis of chemical processes.

Etymology

The etymology of *thermodynamics* has an intricate history. It was first spelled in a hyphenated form as an adjective (*thermo-dynamic*) and from 1854 to 1868 as the noun *thermo-dynamics* to represent the science of generalized heat engines.

American biophysicist Donald Haynie claims that *thermodynamics* was coined in 1840 from the Greek root *therme,* meaning heat and *dynamis,* meaning power. However, this etymology has been cited as unlikely.

Pierre Perrot claims that the term *thermodynamics* was coined by James Joule in 1858 to designate the science of relations between heat and power, however, Joule never used that term, but used instead the term *perfect thermo-dynamic engine* in reference to Thomson's 1849 phraseology.

By 1858, *thermo-dynamics*, as a functional term, was used in William Thomson's paper *An Account of Carnot's Theory of the Motive Power of Heat.*

Branches of Description

The study of thermodynamical systems has developed into several related branches, each using a different fundamental model as a theoretical or experimental basis, or applying the principles to varying types of systems.

Classical Thermodynamics

Classical thermodynamics is the description of the states of thermodynamic systems at near-equilibrium, that uses macroscopic, measurable properties. It is used to model exchanges of energy, work and heat based on the laws of thermodynamics. The qualifier *classical* reflects the fact that it represents the first level of understanding of the subject as it developed in the 19th century and describes the changes of a system in terms of macroscopic empirical (large scale, and measurable) parameters. A microscopic interpretation of these concepts was later provided by the development of *statistical mechanics.*

Statistical Mechanics

Statistical mechanics, also called statistical thermodynamics, emerged with the development of atomic and molecular theories in the late 19th century and early 20th century, and supplemented classical thermodynamics with an interpretation of the microscopic interactions between individual particles or quantum-mechanical states. This field relates the microscopic properties of individual atoms and molecules to the macroscopic, bulk properties of materials that can be observed on the human scale, thereby explaining classical thermodynamics as a natural result of statistics, classical mechanics, and quantum theory at the microscopic level.

Chemical Thermodynamics

Chemical thermodynamics is the study of the interrelation of energy with chemical reactions or with a physical change of state within the confines of the laws of thermodynamics.

Treatment of Equilibrium

Equilibrium thermodynamics is the systematic study of transformations of matter and energy in systems as they approach equilibrium. The word equilibrium implies a state of balance. In an equilibrium state there are no unbalanced potentials, or driving forces, within the system. A central aim in equilibrium thermodynamics is: given a system in a well-defined initial state, subject to accurately specified constraints, to calculate what the state of the system will be once it has reached equilibrium.

Non-equilibrium thermodynamics is a branch of thermodynamics that deals with systems that are not in thermodynamic equilibrium. Most systems found in nature are not in thermodynamic equilibrium because they are not in stationary states, and are continuously and discontinuously subject to flux of matter and energy to and from other systems. The thermodynamic study of non-equilibrium systems requires more general concepts than are dealt with by equilibrium thermodynamics. Many natural systems still today remain beyond the scope of currently known macroscopic thermodynamic methods.

Laws of Thermodynamics

Thermodynamics is principally based on a set of four laws which are universally valid when applied to systems that fall within the constraints implied by each. In the various theoretical descriptions of thermodynamics these laws may be expressed in seemingly differing forms, but the most prominent formulations are the following:

- Zeroth law of thermodynamics: *If two systems are each in thermal equilibrium with a third, they are also in thermal equilibrium with each other.*

This statement implies that thermal equilibrium is an equivalence relation on the set of thermodynamic systems under consideration. Systems are said to be in equilibrium if the small, random exchanges between them (e.g. Brownian motion) do not lead to a net change in energy. This law is tacitly assumed in every measurement of temperature. Thus, if one seeks to decide if two bodies are at the same temperature, it is not necessary to bring them into contact and measure any changes of their observable properties in time. The law provides an empirical definition of temperature and justification for the construction of practical thermometers.

The zeroth law was not initially recognized as a law, as its basis in thermodynamical equilibrium was implied in the other laws. The first, second, and third laws had been explicitly stated prior and found common acceptance in the physics community. Once the importance of the zeroth law for the definition of temperature was realized, it was impracticable to renumber the other laws, hence it was numbered the *zeroth law*.

- First law of thermodynamics: *The internal energy of an isolated system is constant.*

The first law of thermodynamics is an expression of the principle of conservation of energy. It states that energy can be transformed (changed from one form to another), but cannot be created or destroyed.

The first law is usually formulated by saying that the change in the internal energy of a closed thermodynamic system is equal to the difference between the heat supplied to the system and the amount of work done by the system on its surroundings. It is important to note that internal energy is a state of the system whereas heat and work modify the state of the system. In other words, a change of internal energy of a system may be achieved by any combination of heat and work added or removed from the system as long as those total to the change of internal energy. The manner by which a system achieves its internal energy is path independent.

- Second law of thermodynamics: *Heat cannot spontaneously flow from a colder location to a hotter location.*

The second law of thermodynamics is an expression of the universal principle of decay observable in nature. The second law is an observation of the fact that over time, differences in temperature, pressure, and chemical potential tend to even out in a physical system that is isolated from the outside world. Entropy is a measure of how much this process has progressed. The entropy of an isolated system which is not in equilibrium will tend to increase over time, approaching a maximum value at equilibrium.

In classical thermodynamics, the second law is a basic postulate applicable to any system involving heat energy transfer; in statistical thermodynamics, the second law is a consequence of the assumed randomness of molecular chaos. There are many versions of the second law, but they all have the same effect, which is to explain the phenomenon of irreversibility in nature.

- Third law of thermodynamics: *As a system approaches absolute zero, all processes cease and the entropy of the system approaches a minimum value.*

The third law of thermodynamics is a statistical law of nature regarding entropy and the impossibility of reaching absolute zero of temperature. This law provides an absolute reference point for the determination of entropy. The entropy determined relative to this point is the absolute entropy. Alternate definitions are, "the entropy of all systems and of all states of a system is smallest at absolute zero," or equivalently "it is impossible to reach the absolute zero of temperature by any finite number of processes".

Absolute zero, at which all activity would stop if it were possible to happen, is −273.15 °C (degrees Celsius), or −459.67 °F (degrees Fahrenheit) or 0 K (kelvin).

System Models

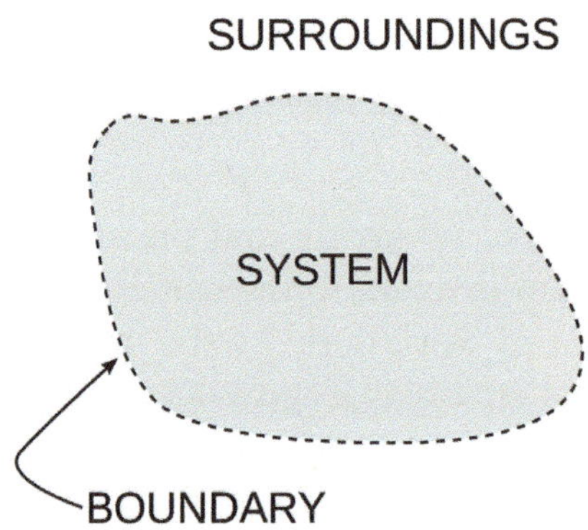

A diagram of a generic thermodynamic system

An important concept in thermodynamics is the thermodynamic system, which is a precisely defined region of the universe under study. Everything in the universe except the system is called the *surroundings*. A system is separated from the remainder of the universe by a *boundary* which may be a physical boundary or notional, but which by convention defines a finite volume. Exchanges of work, heat, or matter between the system and the surroundings take place across this boundary.

In practice, the boundary of a system is simply an imaginary dotted line drawn around a volume within which is going to be a change in the internal energy of that volume. Anything that passes across the boundary that effects a change in the internal energy of the system needs to be accounted for in the energy balance equation. The volume can be the region surrounding a single atom resonating energy, such as Max Planck defined in 1900; it can be a body of steam or air in a steam engine, such as Sadi Carnot defined in 1824; it can be the body of a tropical cyclone, such as Kerry Emanuel theorized in 1986 in the field of atmospheric thermodynamics; it could also be just one nuclide (i.e. a system of quarks) as hypothesized in quantum thermodynamics, or the event horizon of a black hole.

Boundaries are of four types: fixed, movable, real, and imaginary. For example, in an engine, a fixed boundary means the piston is locked at its position, within which a constant volume process might occur. If the piston is allowed to move that boundary is movable while the cylinder and cylinder head boundaries are fixed. For closed systems, boundaries are real while for open systems boundaries are often imaginary. In the case of a jet engine, a fixed imaginary boundary might be assumed at the intake of the engine, fixed boundaries along the surface of the case and a second fixed imaginary boundary across the exhaust nozzle.

Generally, thermodynamics distinguishes three classes of systems, defined in terms of what is allowed to cross their boundaries:

Interactions of thermodynamic systems			
Type of system	Mass flow	Work	Heat
Open	√	√	√
Closed	✕	√	√
Thermally isolated	✕	√	✕
Mechanically isolated	✕	✕	√
Isolated	✕	✕	✕

As time passes in an isolated system, internal differences of pressures, densities, and temperatures tend to even out. A system in which all equalizing processes have gone to completion is said to be in a state of thermodynamic equilibrium.

Once in thermodynamic equilibrium, a system's properties are, by definition, unchanging in time. Systems in equilibrium are much simpler and easier to understand than are systems which are not in equilibrium. Often, when analysing a dynamic thermodynamic process, the simplifying assumption is made that each intermediate state in the process is at equilibrium, producing thermodynamic processes which develop so slowly as to allow each intermediate step to be an equilibrium state and are said to be reversible processes.

States and Processes

When a system is at equilibrium under a given set of conditions, it is said to be in a definite thermodynamic state. The state of the system can be described by a number of state quantities that do not depend on the process by which the system arrived at its state. They are called intensive variables or extensive variables according to how they change when the size of the system changes. The properties of the system can be described by an equation of state which specifies the relationship between these variables. State may be thought of as the instantaneous quantitative description of a system with a set number of variables held constant.

A thermodynamic process may be defined as the energetic evolution of a thermodynamic system proceeding from an initial state to a final state. It can be described by process quantities. Typically, each thermodynamic process is distinguished from other processes in energetic character according to what parameters, such as temperature, pressure, or volume, etc., are held fixed. Furthermore, it is useful to group these processes into pairs, in which each variable held constant is one member of a conjugate pair.

Several commonly studied thermodynamic processes are:

- Adiabatic process: occurs without loss or gain of energy by heat
- Isenthalpic process: occurs at a constant enthalpy
- Isentropic process: a reversible adiabatic process, occurs at a constant entropy
- Isobaric process: occurs at constant pressure
- Isochoric process: occurs at constant volume (also called isometric/isovolumetric)
- Isothermal process: occurs at a constant temperature
- Steady state process: occurs without a change in the internal energy

Instrumentation

There are two types of thermodynamic instruments, the meter and the reservoir. A thermodynamic meter is any device which measures any parameter of a thermodynamic system. In some cases, the thermodynamic parameter is actually defined in terms of an idealized measuring instrument. For example, the zeroth law states that if two bodies are in thermal equilibrium with a third body, they are also in thermal equilibrium with each other. This principle, as noted by James Maxwell in 1872, asserts that it is possible to measure temperature. An idealized thermometer is a sample of an ideal gas at constant pressure. From the ideal gas law $pV=nRT$, the volume of such a sample can be used as an indicator of temperature; in this manner it defines temperature. Although pressure is defined mechanically, a pressure-measuring device, called a barometer may also be constructed from a sample of an ideal gas held at a constant temperature. A calorimeter is a device which is used to measure and define the internal energy of a system.

A thermodynamic reservoir is a system which is so large that its state parameters are not appreciably altered when it is brought into contact with the system of interest. When the reservoir is brought into contact with the system, the system is brought into equilibrium with the reservoir. For example, a pressure reservoir is a system at a particular pressure, which imposes that pressure upon the system to which it is mechanically connected. The Earth's atmosphere is often used as a pressure reservoir. If ocean water is used to cool a power plant, the ocean is often a temperature reservoir in the analysis of the power plant cycle.

Conjugate Variables

The central concept of thermodynamics is that of energy, the ability to do work. By the First Law, the total energy of a system and its surroundings is conserved. Energy may be transferred into a system by heating, compression, or addition of matter, and extracted from a system by cooling, expansion, or extraction of matter. In mechanics, for example, energy transfer equals the product of the force applied to a body and the resulting displacement.

Conjugate variables are pairs of thermodynamic concepts, with the first being akin to a "force" applied to some thermodynamic system, the second being akin to the resulting "displacement," and the product of the two equalling the amount of energy transferred. The common conjugate variables are:

- Pressure-volume (the mechanical parameters);
- Temperature-entropy (thermal parameters);
- Chemical potential-particle number (material parameters).

Potentials

Thermodynamic potentials are different quantitative measures of the stored energy in a system. Potentials are used to measure energy changes in systems as they evolve from an initial state to a final state. The potential used depends on the constraints of the system, such as constant temperature or pressure. For example, the Helmholtz and Gibbs energies are the energies available in a system to do useful work when the temperature and volume or the pressure and temperature are fixed, respectively.

The five most well known potentials are:

Name	Symbol	Formula	Natural variables
Internal energy	U	$\int(TdS - pdV + \sum_i \mu_i dN_i)$	$S, V, \{N_i\}$
Helmholtz free energy	F	$U - TS$	$T, V, \{N_i\}$
Enthalpy	H	$U + pV$	$S, p, \{N_i\}$
Gibbs free energy	G	$U + pV - TS$	$T, p, \{N_i\}$
Landau Potential (Grand potential)	Ω, Φ_G	$U - TS - \sum_i \mu_i N_i$	$T, V, \{\mu_i\}$

where T is the temperature, S the entropy, p the pressure, V the volume, μ the chemical potential, N the number of particles in the system, and i is the count of particles types in the system.

Thermodynamic potentials can be derived from the energy balance equation applied to a thermodynamic system. Other thermodynamic potentials can also be obtained through Legendre transformation.

Applied Fields

- Atmospheric thermodynamics
- Biological thermodynamics
- Black hole thermodynamics
- Chemical thermodynamics
- Classical thermodynamics
- Equilibrium thermodynamics
- Industrial ecology (re: Exergy)
- Maximum entropy thermodynamics
- Non-equilibrium thermodynamics
- Philosophy of thermal and statistical physics
- Psychrometrics
- Quantum thermodynamics
- Statistical thermodynamics
- Thermoeconomics

Fundamental Concepts and Definitions

- In our study of thermodynamics, we will choose a small part of the universe to which we will apply the laws of thermodynamics. We call this subset a SYSTEM.
- The thermodynamic system is analogous to the free body diagram to which we apply the laws of mechanics, (i.e. Newton's Laws of Motion).
- The system is a macroscopically identifiable collection of matter on which we focus our attention (eg: the water kettle or the aircraft engine).
- The rest of the universe outside the system close enough to the system to have some perceptible effect on the system is called the surroundings.
- The surfaces which separates the system from the surroundings are called the

boundaries as shown in figure below (eg: walls of the kettle, the housing of the engine).

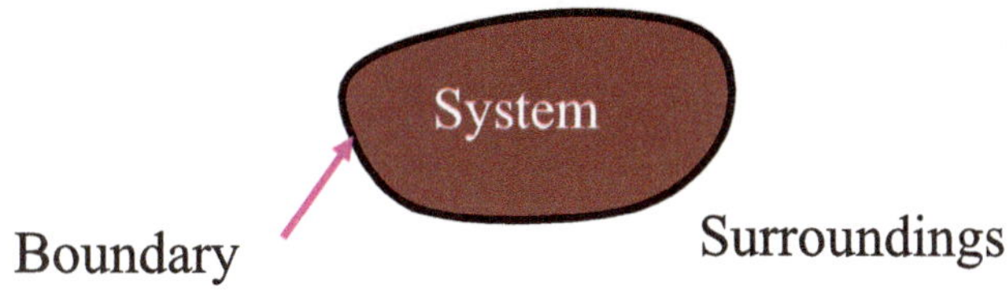

Types of System

- Closed system - in which no mass is permitted to cross the system boundary i.e. we would always consider a system of constant mass.We do permit heat and work to enter or leave but not mass.

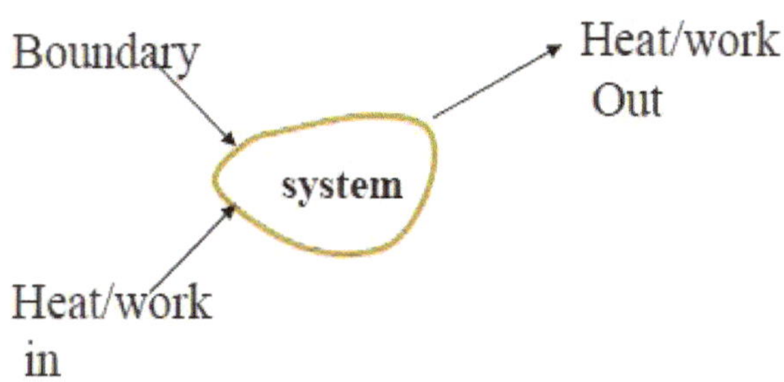

- Open system- in which we permit mass to cross the system boundary in either direction (from the system to surroundings or vice versa). In analysing open systems, we typically look at a specified region of space, and observe what happens at the boundaries of that region.

Most of the Engineering Devices are Open System.

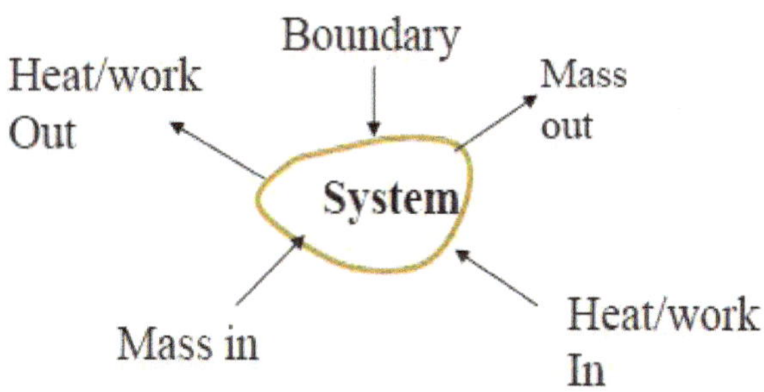

- Isolated System - in which there is no interaction between system and the surroundings. It is of fixed mass and energy, and hence there is no mass and energy transfer across the system boundary.

Thermodynamic System

A thermodynamic system is the material and radiative content of a macroscopic volume in space, that can be adequately described by thermodynamic state variables such as temperature, entropy, internal energy and pressure. Usually, by default, a thermodynamic system is taken to be in its own internal state of thermodynamic equilibrium, as opposed to a non-equilibrium state. The thermodynamic system is always enclosed by walls that separate it from its surroundings; these constrain the system. A thermodynamic system is subject to external interventions called thermodynamic operations; these alter the system's walls or its surroundings; as a result, the system undergoes thermodynamic processes according to the principles of thermodynamics. (This account mainly refers to the simplest kind of thermodynamic system; compositions of simple systems may also be considered.)

The thermodynamic state of a thermodynamic system is its internal state as specified by its state variables. In addition to the state variables, a thermodynamic account also requires a special kind of quantity called a state function, which is a function of the defining state variables. For example, if the state variables are internal energy, volume and mole amounts, that special function is the entropy. These quantities are inter-related by one or more functional relationships called equations of state, and by the system's characteristic equation. Thermodynamics imposes restrictions on the possible equations of state and on the characteristic equation. The restrictions are imposed by the laws of thermodynamics.

According to the permeabilities of the walls of a system, transfers of energy and matter occur between it and its surroundings, which are assumed to be unchanging over time, until a state of thermodynamic equilibrium is attained. The only states considered in equilibrium thermodynamics are equilibrium states. Classical thermodynamics includes equilibrium thermodynamics. It also considers: (a) systems considered in terms of cyclic sequences of processes rather than of states of the system; such were historically important in the conceptual development of the subject; and (b) systems considered in terms of processes described by steady flows; such are important in engineering.

In 1824 Sadi Carnot described a thermodynamic system as the working substance (such as the volume of steam) of any heat engine under study. The very existence of such thermodynamic systems may be considered a fundamental postulate of equilibrium thermodynamics, though it is only rarely cited as a numbered law. According to Bailyn, the commonly rehearsed statement of the zeroth law of thermodynamics is a consequence of this fundamental postulate.

In equilibrium thermodynamics the state variables do not include fluxes because in a state of thermodynamic equilibrium all fluxes have zero values by postulation. Equilibrium thermodynamic processes may of course involve fluxes but these must have ceased by the time a thermodynamic process or operation is complete bringing a system to its eventual thermodynamic state. Non-equilibrium thermodynamics allows its state variables to include non-zero fluxes, that describe transfers of matter or energy or entropy between a system and its surroundings.

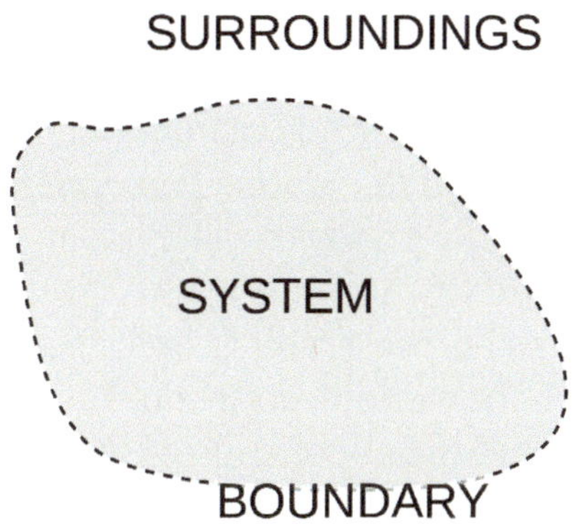

Overview

Thermodynamic equilibrium is characterized by absence of flow of matter or energy. Equilibrium thermodynamics, as a subject in physics, considers macroscopic bodies of matter and energy in states of internal thermodynamic equilibrium. It uses the concept of thermodynamic processes, by which bodies pass from one equilibrium state to another by transfer of matter and energy between them. The term 'thermodynamic system' is used to refer to bodies of matter and energy in the special context of thermodynamics. The possible equilibria between bodies are determined by the physical properties of the walls that separate the bodies. Equilibrium thermodynamics in general does not measure time. Equilibrium thermodynamics is a relatively simple and well settled subject. One reason for this is the existence of a well defined physical quantity called 'the entropy of a body'.

Non-equilibrium thermodynamics, as a subject in physics, considers bodies of matter and energy that are not in states of internal thermodynamic equilibrium, but are usually participating in processes of transfer that are slow enough to allow description in terms of quantities that are closely related to thermodynamic state variables. It is characterized by presence of flows of matter and energy. For this topic, very often the bodies considered have smooth spatial inhomogeneities, so that spatial gradients, for example a temperature gradient, are well enough defined. Thus the description of non-equilibrium thermodynamic systems is a field theory, more complicated than the theory of equilibrium thermodynamics. Non-equilibrium thermodynamics is a growing subject, not an established edifice. In general, it is not possible to find an exactly defined entropy for non-equilibrium problems. For many non-equilibrium thermodynamical problems, an approximately defined quantity called 'time rate of entropy production' is very useful. Non-equilibrium thermodynamics is mostly beyond the scope of the present article.

Another kind of thermodynamic system is considered in engineering. It takes part in a flow process. The account is in terms that approximate, well enough in practice in many cases, equilibrium thermodynamical concepts. This is mostly beyond the scope of the present article, and is set out in other articles, for example the article Flow process.

History

The first to create the concept of a thermodynamic system was the French physicist Sadi Carnot whose 1824 *Reflections on the Motive Power of Fire* studied what he called the *working substance*, e.g., typically a body of water vapor, in steam engines, in regards to the system's ability to do work when heat is applied to it. The working substance could be put in contact with either a heat reservoir (a boiler), a cold reservoir (a stream of cold water), or a piston (to which the working body could do work by pushing on it). In 1850, the German physicist Rudolf Clausius generalized this picture to include the concept of the surroundings, and began referring to the system as a "working body." In his 1850 manuscript *On the Motive Power of Fire*, Clausius wrote:

> "With every change of volume (to the working body) a certain amount work must be done by the gas or upon it, since by its expansion it overcomes an external pressure, and since its compression can be brought about only by an exertion of external pressure. To this excess of work done by the gas or upon it there must correspond, by our principle, a proportional excess of heat consumed or produced, and the gas cannot give up to the "surrounding medium" the same amount of heat as it receives."

The article Carnot heat engine shows the original piston-and-cylinder diagram used by Carnot in discussing his ideal engine; below, we see the Carnot engine as is typically modeled in current use:

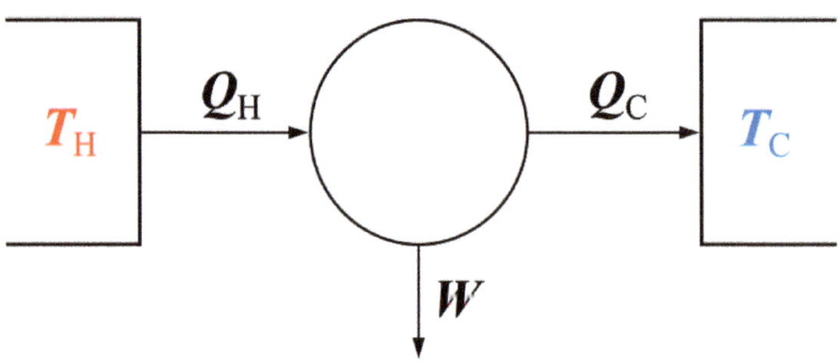

Carnot engine diagram (modern) - where heat flows from a high temperature T_H furnace through the fluid of the "working body" (working substance) and into the cold sink T_C, thus forcing the working substance to do mechanical work W on the surroundings, via cycles of contractions and expansions.

In the diagram shown, the "working body" (system), a term introduced by Clausius in 1850, can be any fluid or vapor body through which heat Q can be introduced or transmitted through to produce work. In 1824, Sadi Carnot, in his famous paper *Reflections on the Motive Power of Fire*, had postulated that the fluid body could be any substance capable of expansion, such as vapor of water, vapor of alcohol, vapor of mercury, a permanent gas, or air, etc. Though, in these early years, engines came in a number of configurations, typically Q_H was supplied by a boiler, wherein water boiled over a furnace; Q_C was typically a stream of cold flowing water in the form of a condenser located on a separate part of the engine. The output work W was the movement of the piston as it turned a crank-arm, which typically turned a pulley to lift water out of flooded salt mines. Carnot defined work as "weight lifted through a height."

Systems in Equilibrium

At thermodynamic equilibrium, a system's properties are, by definition, unchanging in time. Systems in equilibrium are much simpler and easier to understand than systems not in equilibrium. In some cases, when analyzing a thermodynamic process, one can assume that each intermediate state in the process is at equilibrium. This considerably simplifies the analysis.

In isolated systems it is consistently observed that as time goes on internal rearrangements diminish and stable conditions are approached. Pressures and temperatures tend to equalize, and matter arranges itself into one or a few relatively homogeneous phases. A system in which all processes of change have gone practically to completion is considered in a state of thermodynamic equilibrium. The thermodynamic properties of a system in equilibrium are unchanging in time. Equilibrium system states are much easier to describe in a deterministic manner than non-equilibrium states.

For a process to be reversible, each step in the process must be reversible. For a step in a process to be reversible, the system must be in equilibrium throughout the step. That ideal cannot be accomplished in practice because no step can be taken without perturbing the system from equilibrium, but the ideal can be approached by making changes slowly.

Walls

Types of transfers permitted by types of wall			
type of wall	**type of transfer**		
	Matter	**Work**	**Heat**
permeable to matter	✓	✗	✗
permeable to energy but impermeable to matter	✗	✓	✓
adiabatic	✗	✓	✗
adynamic and impermeable to matter	✗	✗	✓
isolating	✗	✗	✗

A system is enclosed by walls that bound it and connect it to its surroundings. Often a wall restricts passage across it by some form of matter or energy, making the connection indirect. Sometimes a wall is no more than an imaginary two-dimensional closed surface through which the connection to the surroundings is direct.

A wall can be fixed (e.g. a constant volume reactor) or moveable (e.g. a piston). For example, in a reciprocating engine, a fixed wall means the piston is locked at its position; then, a constant volume process may occur. In that same engine, a piston may be unlocked and allowed to move in and out. Ideally, a wall may be declared adiabatic, diathermal, impermeable, permeable, or semi-permeable. Actual physical materials that provide walls with such idealized properties are not always readily available.

The system is delimited by walls or boundaries, either actual or notional, across which conserved (such as matter and energy) or unconserved (such as entropy) quantities can pass into and out of the system. The space outside the thermodynamic system is known as the *surroundings*, a *reservoir*, or the *environment*. The properties of the walls determine what transfers can occur. A wall that allows transfer of a quantity is said to be permeable to it, and a thermodynamic system is classified by the permeabilities of its several walls. A transfer between system and surroundings can arise by contact, such as conduction of heat, or by long-range forces such as an electric field in the surroundings.

A system with walls that prevent all transfers is said to be isolated. This is an idealized conception, because in practice some transfer is always possible, for example by gravitational forces. It is an axiom of thermodynamics that an isolated system eventually reaches internal thermodynamic equilibrium, when its state no longer changes with time.

The walls of a closed system allow transfer of energy as heat and as work, but not of matter, between it and its surroundings. The walls of an *open system* allow transfer both of matter and of energy. This scheme of definition of terms is not uniformly used, though it is convenient for some purposes. In particular, some writers use 'closed system' where 'isolated system' is here used.

Anything that passes across the boundary and effects a change in the contents of the system must be accounted for in an appropriate balance equation. The volume can be the region surrounding a single atom resonating energy, such as Max Planck defined in 1900; it can be a body of steam or air in a steam engine, such as Sadi Carnot defined in 1824. It could also be just one nuclide (i.e. a system of quarks) as hypothesized in quantum thermodynamics.

Surroundings

The system is the part of the universe being studied, while the *surroundings* is the remainder of the universe that lies outside the boundaries of the system. It is also known as the *environment*, and the *reservoir*. Depending on the type of system, it may interact with the system by exchanging mass, energy (including heat and work), momentum, electric charge, or other conserved properties. The environment is ignored in analysis of the system, except in regards to these interactions.

Closed System

In a closed system, no mass may be transferred in or out of the system boundaries. The system always contains the same amount of matter, but heat and work can be exchanged across the boundary of the system. Whether a system can exchange heat, work, or both is dependent on the property of its boundary.

- Adiabatic boundary – not allowing any heat exchange: A thermally isolated system
- Rigid boundary – not allowing exchange of work: A mechanically isolated system

One example is fluid being compressed by a piston in a cylinder. Another example of a closed system is a bomb calorimeter, a type of constant-volume calorimeter used in measuring the heat of combustion of a particular reaction. Electrical energy travels across the boundary to produce a spark between the electrodes and initiates combustion. Heat transfer occurs across the boundary after combustion but no mass transfer takes place either way.

Beginning with the first law of thermodynamics for an open system, this is expressed as:

$$\Delta U = Q - W + m_i(h + \frac{1}{2}v^2 + gz)_i - m_e(h + \frac{1}{2}v^2 + gz)_e$$

where *U* is internal energy, *Q* is the heat added to the system, *W* is the work done by the system, and since no mass is transferred in or out of the system, both expressions involving mass flow are zero and the first law of thermodynamics for a closed system is derived. The first law of thermodynamics for a closed system states that the increase of internal energy of the system equals the amount of heat added to the system minus the work done by the system. For infinitesimal changes the first law for closed systems is stated by:

$$dU = \delta Q - \delta W.$$

If the work is due to a volume expansion by d*V* at a pressure *P* then:

$$\delta W = PdV.$$

For a homogeneous system undergoing a reversible process, the second law of thermodynamics reads:

$$\delta Q = TdS$$

where T is the absolute temperature and S is the entropy of the system. With these relations the fundamental thermodynamic relation, used to compute changes in internal energy, is expressed as:

$$dU = TdS - PdV.$$

For a simple system, with only one type of particle (atom or molecule), a closed system amounts to a constant number of particles. However, for systems undergoing a chemical reaction, there may be all sorts of molecules being generated and destroyed by the reaction process. In this case, the fact that the system is closed is expressed by stating that the total number of each elemental atom is conserved, no matter what kind of molecule it may be a part of. Mathematically:

$$\sum_{j=1}^{m} a_{ij} N_j = b_i^0$$

where N_j is the number of j-type molecules, a_{ij} is the number of atoms of element i in molecule j and b_i^o is the total number of atoms of element i in the system, which remains constant, since the system is closed. There is one such equation for each element in the system.

Isolated System

An isolated system is more restrictive than a closed system as it does not interact with its surroundings in any way. Mass and energy remains constant within the system, and no energy or mass transfer takes place across the boundary. As time passes in an isolated system, internal differences in the system tend to even out and pressures and temperatures tend to equalize, as do density differences. A system in which all equalizing processes have gone practically to completion is in a state of thermodynamic equilibrium.

Truly isolated physical systems do not exist in reality (except perhaps for the universe as a whole), because, for example, there is always gravity between a system with mass and masses elsewhere. However, real systems may behave nearly as an isolated system for finite (possibly very long) times. The concept of an isolated system can serve as a useful model approximating many real-world situations. It is an acceptable idealization used in constructing mathematical models of certain natural phenomena.

In the attempt to justify the postulate of entropy increase in the second law of thermodynamics, Boltzmann's H-theorem used equations, which assumed that a system (for example, a gas) was isolated. That is all the mechanical degrees of freedom could be specified, treating the walls simply as mirror boundary conditions. This inevitably led to Loschmidt's paradox. However, if the stochastic behavior of the molecules in actual walls is considered, along with the randomizing effect of the ambient, background thermal radiation, Boltzmann's assumption of molecular chaos can be justified.

The second law of thermodynamics for isolated systems states that the entropy of an isolated system not in equilibrium tends to increase over time, approaching maximum value at equilibrium. Overall, in an isolated system, the internal energy is constant and the entropy can never decrease. A *closed* system's entropy can decrease e.g. when heat is extracted from the system.

It is important to note that isolated systems are not equivalent to closed systems. Closed systems cannot exchange matter with the surroundings, but can exchange energy. Isolated systems can exchange neither matter nor energy with their surroundings, and as such are only theoretical and do not exist in reality (except, possibly, the entire universe).

It is worth noting that 'closed system' is often used in thermodynamics discussions when 'isolated system' would be correct - i.e. there is an assumption that energy does not enter or leave the system.

Selective Transfer of Matter

For a thermodynamic process, the precise physical properties of the walls and surroundings of the system are important, because they determine the possible processes.

An open system has one or several walls that allow transfer of matter. To account for the internal energy of the open system, this requires energy transfer terms in addition to those for heat and work. It also leads to the idea of the chemical potential.

A wall selectively permeable only to a pure substance can put the system in diffusive contact with a reservoir of that pure substance in the surroundings. Then a process is possible in which that pure substance is transferred between system and surroundings. Also, across that wall a contact equilibrium with respect to that substance is possible. By suitable thermodynamic operations, the pure substance reservoir can be dealt with as a closed system. Its internal energy and its entropy can be determined as functions of its temperature, pressure, and mole number.

A thermodynamic operation can render impermeable to matter all system walls other than the contact equilibrium wall for that substance. This allows the definition of an intensive state variable, with respect to a reference state of the surroundings, for that substance. The intensive variable is called the chemical potential; for component substance i it is usually denoted μ_i. The corresponding extensive variable can be the number of moles N_i of the component substance in the system.

For a contact equilibrium across a wall permeable to a substance, the chemical potentials of the substance must be same on either side of the wall. This is part of the nature of thermodynamic equilibrium, and may be regarded as related to the zeroth law of thermodynamics.

Open System

In an open system, matter may pass in and out of some segments of the system boundaries. There may be other segments of the system boundaries that pass heat or work but not matter. Respective account is kept of the transfers of energy across those and any other several boundary segments. In thermodynamic equilibrium, all flows have vanished.

Adiabatic System

Adiabatic system is the one which doesnt allow any heat to be transferred into or out of the system. PV^(gamma) =C equation is only valid for an adiabatic system which is also undergoing a reversible process provided it is a closed system having an ideal gas. If it fails to satisfy any of thes conditions then only dQ=o is true and it cant be represented in an equation like PV^(gamma)=C.

Choice of System and Boundaries

- We must choose the system for each and every problem we work on, so as to obtain best possible information on how it behaves.
- In some cases the choice of the system will be obvious and in some cases not so obvious.
- Important: you must be clear in defining what constitutes your system and make that choice explicit to anyone else who may be reviewing your work.
- The boundaries may be real physical surfaces or they may be imaginary for the convenience of analysis.

eg: If the air in this room is the system,the floor,ceiling and walls constitutes real boundaries.the plane at the open doorway constitutes an imaginary boundary.

- The boundaries may be at rest or in motion.

eg: If we choose a system that has a certain defined quantity of mass (such as gas contained in a piston cylinder device) the boundaries must move in such way that they always enclose that particular quantity of mass if it changes shape or moves from one place to another.

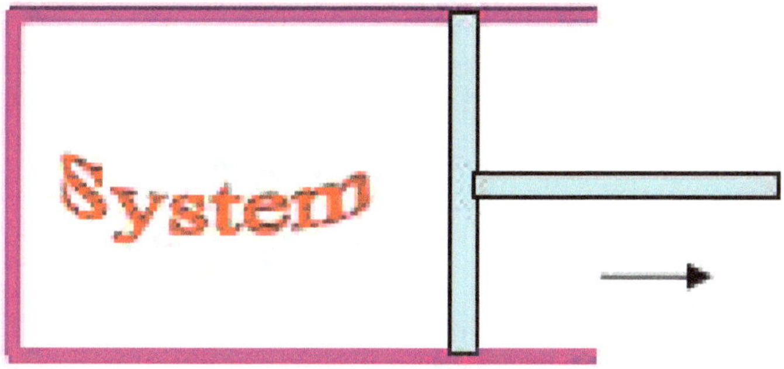

Behavior of matter can be studied by these two approaches.

- In macroscopic approach, certain quantity of matter is considered,without a concern on the events occurring at the molecular level. These effects can be perceived by human senses or measured by instruments.
- eg: pressure, temperature
- In microscopic approach, the effect of molecular motion is Considered.

eg: At microscopic level the pressure of a gas is not constant, the temperature of a gas is a function of the velocity of molecules.

Most microscopic properties cannot be measured with common instruments nor can be perceived by human senses

- It is some characteristic of the system to which some physically meaningful numbers can be assigned without knowing the history behind it.
- These are macroscopic in nature.
- Invariably the properties must enable us to identify the system.
- eg: Anand weighs 72 kg and is 1.75 m tall. We are not concerned how he got to that stage. We are not interested what he ate.

We must choose the most appropriate set of properties.

- For example: Anand weighing 72 kg and being 1.75 m tall may be a useful way of identification for police purposes.
- If he has to work in a company you would say Anand graduated from IIT, Chennai in 1985 in mechanical engineering.
- Anand hails from Mangalore. He has a sister and his father is a poet. He is singer — If you are looking at him as a bridegroom!

- All of them are properties of Anand. But you pick and choose a set of his traits which describe him best for a given situation.
- Similarly, among various properties by which a definition of a thermodynamic system is possible, a situation might warrant giving the smallest number of properties which describe the system best.
- Extensive property:

  whose value depends on the size or extent of the system (upper case letters as the symbols).

  eg: Volume, Mass (V,M).

  If mass is increased, the value of extensive property also increases.
- Intensive property:

  Whose value is independent of the size or extent of the system.

  eg: pressure, temperature (p, T).

  Specific property:
- It is the value of an extensive property per unit mass of system. (lower case letters as symbols) eg: specific volume, density (v, ρ).
- It is a special case of an intensive property.
- Most widely referred properties in thermodynamics:
- Pressure; Volume; Temperature; Entropy; Enthalpy; Internal energy
- State:

  It is the condition of a system as defined by the values of all its properties.

  It gives a complete description of the system.

  Any operation in which one or more properties of a system change is called a change of state.
- Phase:

  It is a quantity of mass that is homogeneous throughout in chemical composition and physical structure.

  e.g. solid, liquid, vapour, gas.

  Phase consisting of more than one phase is known as heterogenous system .

Path and Process

The succession of states passed through during a change of state is called the path of the system. A system is said to go through a proocess if it goes through a series of changes in state. Consequently:

- A system may undergo changes in some or all of its properties.
- A process can be construed to be the locus of changes of state

Processes in thermodynamics are like streets in a city

eg: we have north to south; east to west; roundabouts; crescents

Types of Processes

- As a matter of rule we allow one of the properties to remain a constant during a process.
- Construe as many processes as we can (with a different property kept constant during each of them)
- Complete the cycle by regaining the initial state
    - Isothermal (T)
    - Isobaric (p)
    - Isochoric (v)
    - Isentropic (s)
    - Isenthalpic (h)
    - Isosteric (concentration)
    - Adiabatic (no heat addition or removal

Quasistatic Process

In thermodynamics, a quasi-static process is a thermodynamic process that happens slowly enough for the system to remain in internal equilibrium. An example of this is quasi-static compression, where the volume of a system changes at a rate slow enough to allow the pressure to remain uniform and constant throughout the system.

Any reversible process is necessarily a quasi-static one. However, quasi-static processes involving entropy production are irreversible. An example of a quasi-static process that is not reversible is a compression against a system with a piston subject to friction — although the system is always in thermal equilibrium, the friction ensures the generation of dissipative entropy, which directly goes against the definition of reversible. A notable example of a process that is not even quasi-static is the slow heat exchange between two bodies at two finitely different temperatures, where the heat exchange rate is controlled by an approximately adiabatic partition between the two bodies — in this case, no matter how slowly the process takes place, the states of the two bodies are never infinitesimally close to equilibrium, since thermal equilibrium requires that the two bodies be at the same temperature.

Some ambiguity exists in the literature concerning the distinction between quasi-static and reversible processes, as these are sometimes taken as synonyms. The reason is the theorem that any reversible process is also a quasi-static one, even though the converse is not true. In practical situations, it is essential to differentiate between the two: any engineer would remember to include friction when calculating the dissipative entropy generation, so there are no reversible processes in practice. The definition given above is closer to the intuitive understanding of the word "quasi-" (almost) "static", and remains technically different from reversible processes.

PV-Work in Various Quasi-static Processes

1. Constant pressure: Isobaric processes,

$$W_{1-2} = \int PdV = P(V_2 - V_1)$$

2. Constant volume: Isochoric processes,

$$W_{1-2} = \int PdV = 0$$

3. Constant temperature: Isothermal processes,

$W_{1-2} = \int PdV$, where *P* varies with *V* via $PV = P_1V_1 = C$, , so

$$W_{1-2} = P_1V_1 \ln\frac{V_2}{V_1}$$

4. Polytropic processes,

$$W_{1-2} = \frac{P_1V_1 - P_2V_2}{n-1}$$

Thermodynamic Equilibrium

Thermodynamic equilibrium is an axiomatic concept of thermodynamics. It is an internal state of a single thermodynamic system, or a relation between several thermodynamic systems connected by more or less permeable or impermeable walls. In thermodynamic equilibrium there are no net macroscopic flows of matter or of energy, either within a system or between systems. In a system in its own state of internal thermodynamic equilibrium, no macroscopic change occurs. Systems in mutual thermodynamic equilibrium are simultaneously in mutual thermal, mechanical, chemical, and radiative equilibria. Systems can be in one kind of mutual equilibrium, though not in others. In thermodynamic equilibrium, all kinds of equilibrium hold at once and indefinitely, until disturbed by a thermodynamic operation. In a macroscopic equilibrium, almost or perfectly exactly balanced microscopic exchanges occur; this is the physical explanation of the notion of macroscopic equilibrium.

A thermodynamic system in its own state of internal thermodynamic equilibrium has a spatially uniform temperature. Its intensive properties, other than temperature, may be driven to spatial inhomogeneity by an unchanging long range force field imposed on it by its surroundings.

In non-equilibrium systems, by contrast, there are net flows of matter or energy. If such changes can be triggered to occur in a system in which they are not already occurring, it is said to be in a metastable equilibrium.

Though it is not a widely named law, it is an axiom of thermodynamics that there exist states of thermodynamic equilibrium. The second law of thermodynamics states that when a body of material starts from an equilibrium state, in which portions of it are held at different states by more or less permeable or impermeable partitions, and a thermodynamic operation removes or makes the partitions more permeable and it is isolated, then it spontaneously reaches its own new state of internal thermodynamic equilibrium, and this is accompanied by an increase in the sum of the entropies of the portions.

Overview

Classical thermodynamics deals with states of dynamic equilibrium. The state of a system at thermodynamic equilibrium is the one for which some thermodynamic potential is minimized, or for which the entropy (*S*) is maximized, for specified conditions. One such potential is the Helmholtz free energy (*A*), for a system with surroundings at controlled constant temperature and volume:

$$A = U - TS$$

Another potential, the Gibbs free energy (*G*), is minimized at thermodynamic equilibrium in a system with surroundings at controlled constant temperature and pressure:

$$G = U - TS + PV$$

where *T* denotes the absolute thermodynamic temperature, *P* the pressure, *S* the entropy, *V* the volume, and *U* the internal energy of the system.

Thermodynamic equilibrium is the unique stable stationary state that is approached or eventually reached as the system interacts with its surroundings over a long time. The above-mentioned potentials are mathematically constructed to be the thermodynamic quantities that are minimized under the particular conditions in the specified surroundings.

Conditions for Thermodynamic Equilibrium

- For a completely isolated system, S is maximum at thermodynamic equilibrium.
- For a system with controlled constant temperature and volume, A is minimum at thermodynamic equilibrium.
- For a system with controlled constant temperature and pressure, G is minimum at thermodynamic equilibrium.

The various types of equilibriums are achieved as follows:

- Two systems are in *thermal equilibrium* when their temperatures are the same.
- Two systems are in *mechanical equilibrium* when their pressures are the same.
- Two systems are in *diffusive equilibrium* when their chemical potentials are the same.
- All forces are balanced and there is no significant external driving force.

Relation of Exchange Equilibrium between Systems

Often the surroundings of a thermodynamic system may also be regarded as another thermodynamic system. In this view, one may consider the system and its surroundings as two systems in mutual contact, with long-range forces also linking them. The enclosure of the system is the surface of contiguity or boundary between the two systems. In the thermodynamic formalism, that surface is regarded as having specific properties of permeability. For example, the surface of contiguity may be supposed to be permeable only to heat, allowing energy to transfer only as heat. Then the two systems are said to be in thermal equilibrium when the long-range forces are unchanging in time and the transfer of energy as heat between them has slowed and eventually stopped permanently; this is an example of a contact equilibrium. Other kinds of contact equilibrium are defined by other kinds of specific permeability. When two systems are in contact equilibrium with respect to a particular kind of permeability, they have common values of the intensive variable that belongs to that particular kind of permeability. Examples of such intensive variables are temperature, pressure, chemical potential.

A contact equilibrium may be regarded also as an exchange equilibrium. There is a zero balance of rate of transfer of some quantity between the two systems in contact equilibrium. For example, for a wall permeable only to heat, the rates of diffusion of internal energy as heat between the two systems are equal and opposite. An adiabatic wall between the two systems is 'permeable' only to energy transferred as work; at mechanical equilibrium the rates of transfer of energy as work between them are equal and opposite. If the wall is a simple wall, then the rates of transfer of volume across it are also equal and opposite; and the pressures on either side of it are equal. If the adiabatic wall is more complicated, with a sort of leverage, having an area-ratio, then the pressures of the two systems in exchange equilibrium are in the inverse ratio of the volume exchange ratio; this keeps the zero balance of rates of transfer as work.

A radiative exchange can occur between two otherwise separate systems. Radiative exchange equilibrium prevails when the two systems have the same temperature.

Thermodynamic State of Internal Equilibrium of a System

A collection of matter may be entirely isolated from its surroundings. If it has been left undisturbed for an indefinitely long time, classical thermodynamics postulates that it is in a state in which no changes occur within it, and there are no flows within it. This is a thermodynamic state of internal equilibrium. (This postulate is sometimes, but not often, called the "minus first" law of thermodynamics. One textbook calls it the "zeroth law", remarking that the authors think this more befitting that title than its more customary definition, which apparently was suggested by Fowler.)

Such states are a principal concern in what is known as classical or equilibrium thermodynamics, for they are the only states of the system that are regarded as well defined in that subject. A system in contact equilibrium with another system can by a thermodynamic operation be isolated, and upon the event of isolation, no change occurs in it. A system in a relation of contact equilibrium with another system may thus also be regarded as being in its own state of internal thermodynamic equilibrium.

Multiple Contact Equilibrium

The thermodynamic formalism allows that a system may have contact with several other systems at once, which may or may not also have mutual contact, the contacts having respectively different permeabilities. If these systems are all jointly isolated from the rest of the world, those of them that are in contact then reach respective contact equilibria with one another.

If several systems are free of adiabatic walls between each other, but are jointly isolated from the rest of the world, then they reach a state of multiple contact equilibrium, and they have a common temperature, a total internal energy, and a total entropy. Amongst intensive variables, this is a unique property of temperature. It holds even in the presence of long-range forces. (That is, there is no "force" that can maintain temperature discrepancies.) For example, in a system in thermodynamic equilibrium in a vertical gravitational field, the pressure on the top wall is less than that on the bottom wall, but the temperature is the same everywhere.

A thermodynamic operation may occur as an event restricted to the walls that are within the surroundings, directly affecting neither the walls of contact of the system of interest with its surroundings, nor its interior, and occurring within a definitely limited time. For example, an immovable adiabatic wall may be placed or removed within the surroundings. Consequent upon such an operation restricted to the surroundings, the system may be for a time driven away from its own initial internal state of thermodynamic equilibrium. Then, according to the second law of thermodynamics, the whole undergoes changes and eventually reaches a new and final equilibrium with the surroundings. Following Planck, this consequent train of events is called a natural thermodynamic process. It is allowed in equilibrium thermodynamics just because the initial and final states are of thermodynamic equilibrium, even though during the process there is transient departure from thermodynamic equilibrium, when neither the system nor its surroundings are in well defined states of internal equilibrium. A natural process proceeds at a finite rate for the main part of its course. It is thereby radically different from a fictive quasi-static 'process' that proceeds infinitely slowly throughout its course, and is fictively 'reversible'. Classical thermodynamics allows that even though a process may take a very long time to settle to thermodynamic equilibrium, if the main part of its course is at a finite rate, then it is considered to be natural, and to be subject to the second law of thermodynamics, and thereby irreversible. Engineered machines and artificial devices and manipulations are permitted within the surroundings. The allowance of such operations and devices in the surroundings but not in the system is the reason why Kelvin in one of his statements of the second law of thermodynamics spoke of "inanimate" agency; a system in thermodynamic equilibrium is inanimate.

Otherwise, a thermodynamic operation may directly affect a wall of the system.

It is often convenient to suppose that some of the surrounding subsystems are so much larger than the system that the process can affect the intensive variables only of the surrounding subsystems, and they are then called reservoirs for relevant intensive variables.

Local and Global Equilibrium

It is useful to distinguish between global and local thermodynamic equilibrium. In thermodynamics, exchanges within a system and between the system and the outside are controlled by intensive parameters. As an example, temperature controls heat exchanges. *Global thermodynamic equilibrium* (GTE) means that those intensive parameters are homogeneous throughout the whole system, while *local thermodynamic equilibrium* (LTE) means that those intensive parameters are varying in space and time, but are varying so slowly that, for any point, one can assume thermodynamic equilibrium in some neighborhood about that point.

If the description of the system requires variations in the intensive parameters that are too large, the very assumptions upon which the definitions of these intensive parameters are based will break down, and the system will be in neither global nor local equilibrium. For example, it takes a certain number of collisions for a particle to equilibrate to its surroundings. If the average distance it has moved during these collisions removes it from the neighborhood it is equilibrating to, it will never equilibrate, and there will be no LTE. Temperature is, by definition, proportional to the average internal energy of an equilibrated neighborhood. Since there is no equilibrated neighborhood, the concept of temperature breaks down, and the temperature becomes undefined.

It is important to note that this local equilibrium may apply only to a certain subset of particles in the system. For example, LTE is usually applied only to massive particles. In a radiating gas, the photons being emitted and absorbed by the gas need not be in thermodynamic equilibrium with each other or with the massive particles of the gas in order for LTE to exist. In some cases, it is not considered necessary for free electrons to be in equilibrium with the much more massive atoms or molecules for LTE to exist.

As an example, LTE will exist in a glass of water that contains a melting ice cube. The temperature inside the glass can be defined at any point, but it is colder near the ice cube than far away from it. If energies of the molecules located near a given point are observed, they will be distributed according to the Maxwell–Boltzmann distribution for a certain temperature. If the energies of the molecules located near another point are observed, they will be distributed according to the Maxwell–Boltzmann distribution for another temperature.

Local thermodynamic equilibrium does not require either local or global stationarity. In other words, each small locality need not have a constant temperature. However, it does require that each small locality change slowly enough to practically sustain its local Maxwell–Boltzmann distribution of molecular velocities. A global non-equilibrium state can be stably stationary only if it is maintained by exchanges between the system and the outside. For example, a globally-stable stationary state could be maintained inside the glass of water by continuously adding finely powdered ice into it in order to compensate for the melting, and continuously draining off the meltwater. Natural transport phenomena may lead a system from local to global thermodynamic equilibrium. Going back to our example, the diffusion of heat will lead our glass of water toward global thermodynamic equilibrium, a state in which the temperature of the glass is completely homogeneous.

Reservations

Careful and well informed writers about thermodynamics, in their accounts of thermodynamic equilibrium, often enough make provisos or reservations to their statements. Some writers leave such reservations merely implied or more or less unstated.

For example, one widely cited writer, H. B. Callen writes in this context: “In actuality, few systems are in absolute and true equilibrium." He refers to radioactive processes and remarks that they may take "cosmic times to complete, [and] generally can be ignored". He adds "In practice, the criterion for equilibrium is circular. *Operationally, a system is in an equilibrium state if its properties are consistently described by thermodynamic theory!*"

J.A. Beattie and I. Oppenheim write: "Insistence on a strict interpretation of the definition of equilibrium would rule out the application of thermodynamics to practically all states of real systems."

Another author, cited by Callen as giving a "scholarly and rigorous treatment", and cited by Adkins as having written a "classic text", A.B. Pippard writes in that text: "Given long enough a supercooled vapour will eventually condense, ... The time involved may be so enormous, however, perhaps 10^{100} years or more, ... For most purposes, provided the rapid change is not artificially stimulated, the systems may be regarded as being in equilibrium."

Another author, A. Münster, writes in this context. He observes that thermonuclear processes often occur so slowly that they can be ignored in thermodynamics. He comments: "The concept 'absolute equilibrium' or 'equilibrium with respect to all imaginable processes', has therefore, no physical significance." He therefore states that: "... we can consider an equilibrium only with respect to specified processes and defined experimental conditions."

According to L. Tisza: "... in the discussion of phenomena near absolute zero. The absolute predictions of the classical theory become particularly vague because the occurrence of frozen-in nonequilibrium states is very common."

Definitions

The most general kind of thermodynamic equilibrium of a system is through contact with the surroundings that allows simultaneous passages of all chemical substances and all kinds of energy. A system in thermodynamic equilibrium may move with uniform acceleration through space but must not change its shape or size while doing so; thus it is defined by a rigid volume in space. It may lie within external fields of force, determined by external factors of far greater extent than the system itself, so that events within the system cannot in an appreciable amount affect the external fields of force. The system can be in thermodynamic equilibrium only if the external force fields are uniform, and are determining its uniform acceleration, or if it lies in a non-uniform force field but is held stationary there by local forces, such as mechanical pressures, on its surface.

Thermodynamic equilibrium is a primitive notion of the theory of thermodynamics. According to P.M. Morse: "It should be emphasized that the fact that there are thermodynamic states, ..., and the fact that there are thermodynamic variables which are uniquely specified by the equilibrium state ... are *not* conclusions deduced logically from some philosophical first principles. They are conclusions ineluctably drawn from more than two centuries of experiments." This means that thermodynamic equilibrium is not to be defined solely in terms of other theoretical concepts of thermodynamics. M. Bailyn proposes a fundamental law of thermodynamics that defines and postulates the existence of states of thermodynamic equilibrium.

Textbook definitions of thermodynamic equilibrium are often stated carefully, with some reservation or other.

For example, A. Münster writes: "An isolated system is in thermodynamic equilibrium when, in the system, no changes of state are occurring at a measurable rate." There are two reservations stated here; the system is isolated; any changes of state are immeasurably slow. He discusses the second proviso by giving an account of a mixture oxygen and hydrogen at room temperature in the absence of a catalyst. Münster points out that a thermodynamic equilibrium state is described by fewer macroscopic variables than is any other state of a given system. This is partly, but not entirely, because all flows within and through the system are zero.

R. Haase's presentation of thermodynamics does not start with a restriction to thermodynamic equilibrium because he intends to allow for non-equilibrium thermodynamics. He considers an arbitrary system with time invariant properties. He tests it for thermodynamic equilibrium by cutting it off from all external influences, except external force fields. If after insulation, nothing changes, he says that the system was in *equilibrium*.

In a section headed "Thermodynamic equilibrium", H.B. Callen defines equilibrium states in a paragraph. He points out that they "are determined by intrinsic factors" within the system. They are "terminal states", towards which the systems evolve, over time, which may occur with "glacial slowness". This statement does not explicitly say that for thermodynamic equilibrium, the system must be isolated; Callen does not spell out what he means by the words "intrinsic factors".

Another textbook writer, C.J. Adkins, explicitly allows thermodynamic equilibrium to occur in a system which is not isolated. His system is, however, closed with respect to transfer of matter. He writes: "In general, the approach to thermodynamic equilibrium will involve both thermal and work-like interactions with the surroundings." He distinguishes such thermodynamic equilibrium from thermal equilibrium, in which only thermal contact is mediating transfer of energy.

Another textbook author, J.R. Partington, writes: "(i) *An equilibrium state is one which is independent of time.*" But, referring to systems "which are only apparently in equilibrium", he adds : "Such systems are in states of ″false equilibrium.″" Partington's statement does not explicitly state that the equilibrium refers to an isolated system. Like Münster, Partington also refers to the mixture of oxygen and hydrogen. He adds a proviso that "In a true equilibrium state, the smallest change of any external condition which influences the state will produce a small change of state ..." This proviso means that thermodynamic equilibrium must be stable against small perturbations; this requirement is essential for the strict meaning of thermodynamic equilibrium.

A student textbook by F.H. Crawford has a section headed "Thermodynamic Equilibrium". It distinguishes several drivers of flows, and then says: "These are examples of the apparently universal tendency of isolated systems toward a state of complete mechanical, thermal, chemical, and electrical—or, in a single word, *thermodynamic—equilibrium.*"

A monograph on classical thermodynamics by H.A. Buchdahl considers the "equilibrium of a thermodynamic system", without actually writing the phrase "thermodynamic equilibrium". Referring to systems closed to exchange of matter, Buchdahl writes: "If a system is in a terminal condition which is properly static, it will be said to be in *equilibrium.*" Buchdahl's monograph also discusses amorphous glass, for the purposes of thermodynamic description. It states: "More precisely, the glass may be regarded as being *in equilibrium* so long as experimental tests show that 'slow' transitions are in effect reversible." It is not customary to make this proviso part of the definition of thermodynamic equilibrium, but the converse is usually assumed: that if a body in thermodynamic equilibrium is subject to a sufficiently slow process, that process may be considered to be sufficiently nearly reversible, and the body remains sufficiently nearly in thermodynamic equilibrium during the process.

A. Münster carefully extends his definition of thermodynamic equilibrium for isolated systems by introducing a concept of *contact equilibrium*. This specifies particular processes that are allowed when considering thermodynamic equilibrium for non-isolated systems, with special concern for open systems, which may gain or lose matter from or to their surroundings. A contact equilibrium is between the system of interest and a system in the surroundings, brought into contact with the system of interest, the contact being through a special kind of wall; for the rest, the whole joint system is isolated. Walls of this special kind were also considered by C. Carathéodory, and are mentioned by other writers also. They are selectively permeable. They may be permeable only to mechanical work, or only to heat, or only to some particular chemical substance. Each contact equilibrium defines an intensive parameter; for example, a wall permeable only to heat defines an empirical temperature. A contact equilibrium can exist for each chemical constituent of the system of interest. In a contact equilibrium, despite the possible exchange through the selectively permeable wall, the system of interest is changeless, as if it were in isolated thermodynamic equilibrium. This scheme follows the general rule that "... we can consider an equilibrium only with respect to specified processes and defined experimental conditions." Thermodynamic equilibrium for an open system means that, with respect to every relevant kind of selectively permeable wall, contact equilibrium exists when the respective intensive parameters of the system and surroundings are equal. This definition does not consider the most general kind of thermodynamic equilibrium, which is through unselective contacts. This definition does not simply state that no current of matter or energy exists in the interior or at the boundaries; but it is compatible with the following definition, which does so state.

M. Zemansky also distinguishes mechanical, chemical, and thermal equilibrium. He then writes: “When the conditions for all three types of equilibrium are satisfied, the system is said to be in a state of thermodynamic equilibrium”.

P.M. Morse writes that thermodynamics is concerned with “*states of thermodynamic equilibrium*”. He also uses the phrase “thermal equilibrium” while discussing transfer of energy as heat between a body and a heat reservoir in its surroundings, though not explicitly defining a special term ‘thermal equilibrium’.

J.R. Waldram writes of “a definite thermodynamic state”. He defines the term “thermal equilibrium” for a system “when its observables have ceased to change over time”. But shortly below that definition he writes of a piece of glass that has not yet reached its “*full* thermodynamic equilibrium state”.

Considering equilibrium states, M. Bailyn writes: “Each intensive variable has its own type of equilibrium.” He then defines thermal equilibrium, mechanical equilibrium, and material equilibrium. Accordingly, he writes: “If all the intensive variables become uniform, *thermodynamic equilibrium* is said to exist.” He is not here considering the presence of an external force field.

J.G. Kirkwood and I. Oppenheim define thermodynamic equilibrium as follows: “A system is in a state of *thermodynamic equilibrium* if, during the time period allotted for experimentation, (a) its intensive properties are independent of time and (b) no current of matter or energy exists in its interior or at its boundaries with the surroundings.” It is evident that they are not restricting the definition to isolated or to closed systems. They do not discuss the possibility of changes that occur with “glacial slowness”, and proceed beyond the time period allotted for experimentation. They note that for two systems in contact, there exists a small subclass of intensive properties such that if all those of that small subclass are respectively equal, then all respective intensive properties are equal. States of thermodynamic equilibrium may be defined by this subclass, provided some other conditions are satisfied.

Characteristics of a State of Internal Thermodynamic Equilibrium

Homogeneity in the Absence of External Forces

A thermodynamic system consisting of a single phase in the absence of external forces, in its own internal thermodynamic equilibrium, is homogeneous. This means that the material in any small volume element of the system can be interchanged with the material of any other geometrically congruent volume element of the system, and the effect is to leave the system thermodynamically unchanged. In general, a strong external force field makes a system of a single phase in its own internal thermodynamic equilibrium inhomogeneous with respect to some intensive variables. For example, a relatively dense component of a mixture can be concentrated by centrifugation.

Uniform Temperature

Such equilibrium inhomogeneity, induced by external forces, does not occur for the intensive variable temperature. According to E.A. Guggenheim, "The most important conception of thermodynamics is temperature." Planck introduces his treatise with a brief account of heat and temperature and thermal equilibrium, and then announces: "In the following we shall deal chiefly with homogeneous, isotropic bodies of any form, possessing throughout their substance the same temperature and density, and subject to a uniform pressure acting everywhere perpendicular to the surface." As did Carathéodory, Planck was setting aside surface effects and external fields and anisotropic crystals. Though referring to temperature, Planck did not there explicitly refer to the concept of thermodynamic equilibrium. In contrast, Carathéodory's scheme of presentation of classical thermodynamics for closed systems postulates the concept of an "equilibrium state" following Gibbs (Gibbs speaks routinely of a "thermodynamic state"), though not explicitly using the phrase 'thermodynamic equilibrium', nor explicitly postulating the existence of a temperature to define it.

The temperature within a system in thermodynamic equilibrium is uniform in space as well as in time. In a system in its own state of internal thermodynamic equilibrium, there are no net internal macroscopic flows. In particular, this means that all local parts of the system are in mutual radiative exchange equilibrium. This means that the temperature of the system is spatially uniform. This is so in all cases, including those of non-uniform external force fields. For an externally imposed gravitational field, this may be proved in macroscopic thermodynamic terms, by the calculus of variations, using the method of Langrangian multipliers. Considerations of kinetic theory or statistical mechanics also support this statement.

In order that a system may be in its own internal state of thermodynamic equilibrium, it is of course necessary, but not sufficient, that it be in its own internal state of thermal equilibrium; it is possible for a system to reach internal mechanical equilibrium before it reaches internal thermal equilibrium.

Number of Real Variables Needed for Specification

In his exposition of his scheme of closed system equilibrium thermodynamics, C. Carathéodory initially postulates that experiment reveals that a definite number of real variables define the states that are the points of the manifold of equilibria. In the words of Prigogine and Defay (1945): "It is a matter of experience that when we have specified a certain number of macroscopic properties of a system, then all the other properties are fixed." As noted above, according to A. Münster, the number of variables needed to define a thermodynamic equilibrium is the least for any state of a given isolated system.

As noted above, J.G. Kirkwood and I. Oppenheim point out that a state of thermodynamic equilibrium may be defined by a special subclass of intensive variables, with a definite number of members in that subclass.

If the thermodynamic equilibrium lies in an external force field, it is only the temperature that can in general be expected to be spatially uniform. Intensive variables other than temperature will in general be non-uniform if the external force field is non-zero. In such a case, in general, additional variables are needed to describe the spatial non-uniformity.

Stability Against Small Perturbations

As noted above, J.R. Partington points out that a state of thermodynamic equilibrium is stable against small transient perturbations. Without this condition, in general, experiments intended to study systems in thermodynamic equilibrium are in severe difficulties.

Approach to Thermodynamic Equilibrium within an Isolated System

When a body of material starts from a non-equilibrium state of inhomogeneity or chemical non-equilibrium, and is then isolated, it spontaneously evolves towards its own internal state of thermodynamic equilibrium. It is not necessary that all aspects of internal thermodynamic equilibrium be reached simultaneously; some can be established before others. For example, in many cases of such evolution, internal mechanical equilibrium is established much more rapidly than the other aspects of the eventual thermodynamic equilibrium. Another example is that, in many cases of such evolution, thermal equilibrium is reached much more rapidly than chemical equilibrium.

Fluctuations within an Isolated System in its Own Internal Thermodynamic Equilibrium

In an isolated system, thermodynamic equilibrium by definition persists over an indefinitely long time. In classical physics it is often convenient to ignore the effects of measurement and this is assumed in the present account.

To consider the notion of fluctuations in an isolated thermodynamic system, a convenient example is a system specified by its extensive state variables, internal energy, volume, and mass composition. By definition they are time-invariant. By definition, they combine with time-invariant nominal values of their conjugate intensive functions of state, inverse temperature, pressure divided by temperature, and the chemical potentials divided by temperature, so as to exactly obey the laws of thermodynamics. But the laws of thermodynamics, combined with the values of the specifying extensive variables of state, are not sufficient to provide knowledge of those nominal values.

It may be admitted that on repeated measurement of those conjugate intensive functions of state, they are found to have slightly different values from time to time. Such variability is regarded as due to internal fluctuations. The different measured values average to their nominal values.

If the system is truly macroscopic as postulated by classical thermodynamics, then the fluctuations are too small to detect macroscopically. This is called the thermodynamic limit. In effect, the molecular nature of matter and the quantal nature of momentum transfer have vanished from sight, too small. According to Buchdahl: "... there is no place within the strictly phenomenological theory for the idea of fluctuations about equilibrium."

If the system is repeatedly subdivided, eventually a system is produced that is small enough to exhibit obvious fluctuations. This is a mesoscopic level of investigation. The fluctuations are then directly dependent on the natures of the various walls of the system. The precise choice of independent state variables is then important. At this stage, statistical features of the laws of thermodynamics become apparent.

If the mesoscopic system is further repeatedly divided, eventually a microscopic system is produced. Then the molecular character of matter and the quantal nature of momentum transfer become important in the processes of fluctuation. One has left the realm of classical or macroscopic thermodynamics, and one needs quantum statistical mechanics. The fluctuations can become relatively dominant, and questions of measurement become important.

The statement that 'the system is its own internal thermodynamic equilibrium' may be taken to mean that 'indefinitely many such measurements have been taken from time to time, with no trend in time in the various measured values'. Thus the statement, that 'a system is in its own internal thermodynamic equilibrium, with stated nominal values of its functions of state conjugate to its specifying state variables', is far far more informative than a statement that 'a set of single simultaneous measurements of those functions of state have those same values'. This is because the single measurements might have been made during a slight fluctuation, away from another set of nominal values of those conjugate intensive functions of state, that is due to unknown and different constitutive properties. A single measurement cannot tell whether that might be so, unless there is also knowledge of the nominal values that belong to the equilibrium state.

Thermal Equilibrium

An explicit distinction between 'thermal equilibrium' and 'thermodynamic equilibrium' is made by B. C. Eu. He considers two systems in thermal contact, one a thermometer, the other a system in which there are occurring several irreversible processes, entailing non-zero fluxes; the two systems are separated by a wall permeable only to heat. He considers the case in which, over the time scale of interest, it happens that both the thermometer reading and the irreversible processes are steady. Then there is thermal equilibrium without thermodynamic equilibrium. Eu proposes consequently that the zeroth law of thermodynamics can be considered to apply even when thermodynamic equilibrium is not present; also he proposes that if changes are occurring so fast that a steady temperature cannot be defined, then "it is no longer possible to describe the process by means of a thermodynamic formalism. In other words, thermodynamics has no meaning for such a process." This illustrates the importance for thermodynamics of the concept of temperature.

Thermal equilibrium is achieved when two systems in thermal contact with each other cease to have a net exchange of energy. It follows that if two systems are in thermal equilibrium, then their temperatures are the same.

Thermal equilibrium occurs when a system's macroscopic thermal observables have ceased to change with time. For example, an ideal gas whose distribution function has stabilised to a specific Maxwell–Boltzmann distribution would be in thermal equilibrium. This outcome allows a single temperature and pressure to be attributed to the whole system. For an isolated body, it is quite possible for mechanical equilibrium to be reached before thermal equilibrium is reached, but eventually, all aspects of equilibrium, including thermal equilibrium, are necessary for thermodynamic equilibrium.

Non-equilibrium

A system's internal state of thermodynamic equilibrium should be distinguished from a "stationary state" in which thermodynamic parameters are unchanging in time but the system is not isolated, so that there are, into and out of the system, non-zero macroscopic fluxes which are constant in time.

Non-equilibrium thermodynamics is a branch of thermodynamics that deals with systems that are not in thermodynamic equilibrium. Most systems found in nature are not in thermodynamic equilibrium because they are changing or can be triggered to change over time, and are continuously and discontinuously subject to flux of matter and energy to and from other systems. The thermodynamic study of non-equilibrium systems requires more general concepts than are dealt with by equilibrium thermodynamics. Many natural systems still today remain beyond the scope of currently known macroscopic thermodynamic methods.

Non-equilibrium Thermodynamics

Non-equilibrium thermodynamics is a branch of thermodynamics that deals with physical systems that are not in thermodynamic equilibrium but can be described in terms of variables (non-equilibrium state variables) that represent an extrapolation of the variables used to specify the system in thermodynamic equilibrium. Non-equilibrium thermodynamics is concerned with transport processes and with the rates of chemical reactions. It relies on what may be thought of as more or less nearness to thermodynamic equilibrium. Non-equilibrium thermodynamics is a work in progress, not an established edifice. This article will try to sketch some approaches to it and some concepts important for it.

Almost all systems found in nature are not in thermodynamic equilibrium; for they are changing or can be triggered to change over time, and are continuously and discontinuously subject to flux of matter and energy to and from other systems and to chemical reactions. Some systems and processes are, however, in a useful sense, near enough to thermodynamic equilibrium to allow description with useful accuracy by currently known non-equilibrium thermodynamics. Nevertheless, many natural systems and processes will always remain far beyond the scope of non-equilibrium thermodynamic methods. This is because of the very small size of atoms, as compared with macroscopic systems.

The thermodynamic study of non-equilibrium systems requires more general concepts than are dealt with by equilibrium thermodynamics. One fundamental difference between equilibrium thermodynamics and non-equilibrium thermodynamics lies in the behaviour of inhomogeneous systems, which require for their study knowledge of rates of reaction which are not considered in equilibrium thermodynamics of homogeneous systems. Another fundamental and very important difference is the difficulty or impossibility in defining entropy at an instant of time in macroscopic terms for systems not in thermodynamic equilibrium.

Scope of Non-equilibrium Thermodynamics

Difference between Equilibrium and Non-equilibrium Thermodynamics

A profound difference separates equilibrium from non-equilibrium thermodynamics. Equilibrium thermodynamics ignores the time-courses of physical processes. In contrast, non-equilibrium thermodynamics attempts to describe their time-courses in continuous detail.

Equilibrium thermodynamics restricts its considerations to processes that have initial and final states of thermodynamic equilibrium; the time-courses of processes are deliberately ignored. Consequently, equilibrium thermodynamics allows processes that pass through states far from thermodynamic equilibrium, that cannot be described even by the variables admitted for non-equilibrium thermodynamics, such as time rates of change of temperature and pressure. For example, in equilibrium thermodynamics, a process is allowed to include even a violent explosion that cannot be described by non-equilibrium thermodynamics. Equilibrium thermodynamics does, however, for theoretical development, use the idealized concept of the "quasi-static process". A quasi-static process is a conceptual (timeless and physically impossible) smooth mathematical passage along a continuous path of states of thermodynamic equilibrium. It is an exercise in differential geometry rather than a process that could occur in actuality.

Non-equilibrium thermodynamics, on the other hand, attempting to describe continuous time-courses, need its state variables to have a very close connection with those of equilibrium thermodynamics. This profoundly restricts the scope of non-equilibrium thermodynamics, and places heavy demands on its conceptual framework.

Non-equilibrium State Variables

The suitable relationship that defines non-equilibrium thermodynamic state variables is as follows. On occasions when the system happens to be in states that are sufficiently close to thermodynamic equilibrium, non-equilibrium state variables are such that they can be measured locally with sufficient accuracy by the same techniques as are used to measure thermodynamic state variables, or by corresponding time and space derivatives, including fluxes of matter and energy. In general, non-equilibrium thermodynamic systems are spatially and temporally non-uniform, but their non-uniformity still has a sufficient degree of smoothness to support the existence of suitable time and space derivatives of non-equilibrium state variables. Because of the spatial non-uniformity, non-equilibrium state variables that correspond to extensive thermodynamic state variables have to be defined as spatial densities of the corresponding extensive equilibrium state variables. On occasions when the system is sufficiently close to thermodynamic equilibrium, intensive non-equilibrium state variables, for example temperature and pressure, correspond closely with equilibrium state variables. It is necessary that measuring probes be small enough, and rapidly enough responding, to capture relevant non-uniformity. Further, the non-equilibrium state variables are required to be mathematically functionally related to one another in ways that suitably resemble corresponding relations between equilibrium thermodynamic state variables. In reality, these requirements are very demanding, and it may be difficult or practically, or even theoretically, impossible to satisfy them. This is part of why non-equilibrium thermodynamics is a work in progress.

Overview

Non-equilibrium thermodynamics is a work in progress, not an established edifice. This article will try to sketch some approaches to it and some concepts important for it.

Some concepts of particular importance for non-equilibrium thermodynamics include time rate of dissipation of energy (Rayleigh 1873, Onsager 1931, also), time rate of entropy production (Onsager 1931), thermodynamic fields, dissipative structure, and non-linear dynamical structure.

One problem of interest is the thermodynamic study of non-equilibrium steady states, in which entropy production and some flows are non-zero, but there is no time variation of physical variables.

One initial approach to non-equilibrium thermodynamics is sometimes called 'classical irreversible thermodynamics'. There are other approaches to non-equilibrium thermodynamics, for example extended irreversible thermodynamics, and generalized thermodynamics, but they are hardly touched on in the present article.

Quasi-radiationless non-equilibrium Thermodynamics of Matter in laboratory Conditions

According to Wildt, current versions of non-equilibrium thermodynamics ignore radiant heat; they can do so because they refer to laboratory quantities of matter under laboratory conditions with temperatures well below those of stars. At laboratory temperatures, in laboratory quantities of matter, thermal radiation is weak and can be practically nearly ignored. But, for example, atmospheric physics is concerned with large amounts of matter, occupying cubic kilometers, that, taken as a whole, are not within the range of laboratory quantities; then thermal radiation cannot be ignored.

Local Equilibrium Thermodynamics

The terms 'classical irreversible thermodynamics' and 'local equilibrium thermodynamics' are sometimes used to refer to a version of non-equilibrium thermodynamics that demands certain simplifying assumptions, as follows. The assumptions have the effect of making each very small volume element of the system effectively homogeneous, or well-mixed, or without an effective spatial structure, and without kinetic energy of bulk flow or of diffusive flux. Even within the thought-frame of classical irreversible thermodynamics, care is needed in choosing the independent variables for systems. In some writings, it is assumed that the intensive variables of equilibrium thermodynamics are sufficient as the independent variables for the task (such variables are considered to have no 'memory', and do not show hysteresis); in particular, local flow intensive variables are not admitted as independent variables; local flows are considered as dependent on quasi-static local intensive variables.

Also it is assumed that the local entropy density is the same function of the other local intensive variables as in equilibrium; this is called the local thermodynamic equilibrium assumption. Radiation is ignored because it is transfer of energy between regions, which can be remote from one another. In the classical irreversible thermodynamic approach, there is allowed very small spatial variation, from very small volume element to adjacent very small volume element, but it is assumed that the global entropy of the system can be found by simple spatial integration of the local entropy density; this means that spatial structure cannot contribute as it properly should to the global entropy assessment for the system. This approach assumes spatial and temporal continuity and even differentiability of locally defined intensive variables such as temperature and internal energy density. All of these are very stringent demands. Consequently, this approach can deal with only a very limited range of phenomena. This approach is nevertheless valuable because it can deal well with some macroscopically observable phenomena.

In other writings, local flow variables are considered; these might be considered as classical by analogy with the time-invariant long-term time-averages of flows produced by endlessly repeated cyclic processes; examples with flows are in the thermoelectric phenomena known as the Seebeck and the Peltier effects, considered by Kelvin in the nineteenth century and by Onsager in the twentieth. These effects occur at metal junctions, which were originally effectively treated as two-dimensional surfaces, with no spatial volume, and no spatial variation.

Local Equilibrium Thermodynamics with Materials with "Memory"

A further extension of local equilibrium thermodynamics is to allow that materials may have "memory", so that their constitutive equations depend not only on present values but also on past values of local equilibrium variables. Thus time comes into the picture more deeply than for time-dependent local equilibrium thermodynamics with memoryless materials, but fluxes are not independent variables of state.

Extended Irreversible Thermodynamics

Extended irreversible thermodynamics is a branch of non-equilibrium thermodynamics that goes outside the restriction to the local equilibrium hypothesis. The space of state variables is enlarged by including the fluxes of mass, momentum and energy and eventually higher order fluxes. The formalism is well-suited for describing high-frequency processes and small-length scales materials.

Basic Concepts

There are many examples of stationary non-equilibrium systems, some very simple, like a system confined between two thermostats at different temperatures or the ordinary Couette flow, a fluid enclosed between two flat walls moving in opposite directions and defining non-equilibrium conditions at the walls. Laser action is also a non-equilibrium process, but it depends on departure from local thermodynamic equilibrium and is thus beyond the scope of classical irreversible thermodynamics; here a strong temperature difference is maintained between two molecular degrees of freedom (with molecular laser, vibrational and rotational molecular motion), the requirement for two component 'temperatures' in the one small region of space, precluding local thermodynamic equilibrium, which demands that only one temperature be needed. Damping of acoustic perturbations or shock waves are non-stationary non-equilibrium processes. Driven complex fluids, turbulent systems and glasses are other examples of non-equilibrium systems.

The mechanics of macroscopic systems depends on a number of extensive quantities. It should be stressed that all systems are permanently interacting with their surroundings, thereby causing unavoidable fluctuations of extensive quantities. Equilibrium conditions of thermodynamic systems are related to the maximum property of the entropy. If the only extensive quantity that is allowed to fluctuate is the internal energy, all the other ones being kept strictly constant, the temperature of the system is measurable and meaningful. The system's properties are then most conveniently described using the thermodynamic potential Helmholtz free energy ($A = U - TS$), a Legendre transformation of the energy. If, next to fluctuations of the energy, the macroscopic dimensions (volume) of the system are left fluctuating, we use the Gibbs free energy ($G = U + PV - TS$), where the system's properties are determined both by the temperature and by the pressure.

Non-equilibrium systems are much more complex and they may undergo fluctuations of more extensive quantities. The boundary conditions impose on them particular intensive variables, like temperature gradients or distorted collective motions (shear motions, vortices, etc.), often called thermodynamic forces. If free energies are very useful in equilibrium thermodynamics, it must be stressed that there is no general law defining stationary non-equilibrium properties of the energy as is the second law of thermodynamics for the entropy in equilibrium thermodynamics. That is why in such cases a more generalized Legendre transformation should be considered. This is the extended Massieu potential. By definition, the entropy (S) is a function of the collection of extensive quantities E_i. Each extensive quantity has a conjugate intensive variable I_i (a restricted definition of intensive variable is used here by comparison to the definition given in) so that:

$$I_i = \partial S / \partial E_i.$$

We then define the extended Massieu function as follows:

$$k_b M = S - \sum_i (I_i E_i),$$

where k_b is Boltzmann's constant, whence

$$k_b\, dM = \sum_i (E_i\, dI_i).$$

The independent variables are the intensities.

Intensities are global values, valid for the system as a whole. When boundaries impose to the system different local conditions, (e.g. temperature differences), there are intensive variables representing the average value and others representing gradients or higher moments. The latter are the thermodynamic forces driving fluxes of extensive properties through the system.

It may be shown that the Legendre transformation changes the maximum condition of the entropy (valid at equilibrium) in a minimum condition of the extended Massieu function for stationary states, no matter whether at equilibrium or not.

Stationary States, Fluctuations, and Stability

In thermodynamics one is often interested in a stationary state of a process, allowing that the stationary state include the occurrence of unpredictable and experimentally unreproducible fluctuations in the state of the system. The fluctuations are due to the system's internal sub-processes and to exchange of matter or energy with the system's surroundings that create the constraints that define the process.

If the stationary state of the process is stable, then the unreproducible fluctuations involve local transient decreases of entropy. The reproducible response of the system is then to increase the entropy back to its maximum by irreversible processes: the fluctuation cannot be reproduced with a significant level of probability. Fluctuations about stable stationary states are extremely small except near critical points (Kondepudi and Prigogine 1998, page 323). The stable stationary state has a local maximum of entropy and is locally the most reproducible state of the system. There are theorems about the irreversible dissipation of fluctuations. Here 'local' means local with respect to the abstract space of thermodynamic coordinates of state of the system.

If the stationary state is unstable, then any fluctuation will almost surely trigger the virtually explosive departure of the system from the unstable stationary state. This can be accompanied by increased export of entropy.

Local Thermodynamic Equilibrium

The scope of present-day non-equilibrium thermodynamics does not cover all physical processes. A condition for the validity of many studies in non-equilibrium thermodynamics of matter is that they deal with what is known as *local thermodynamic equilibrium*.

Local Thermodynamic Equilibrium of Ponderable Matter

Local thermodynamic equilibrium of matter means that conceptually, for study and analysis, the system can be spatially and temporally divided into 'cells' or 'micro-phases' of small (infinitesimal) size, in which classical thermodynamical equilibrium conditions for matter are fulfilled to good approximation. These conditions are unfulfilled, for example, in very rarefied gases, in which molecular collisions are infrequent; and in the boundary layers of a star, where radiation is passing energy to space; and for interacting fermions at very low temperature, where dissipative processes become ineffective. When these 'cells' are defined, one admits that matter and energy may pass freely between contiguous 'cells', slowly enough to leave the 'cells' in their respective individual local thermodynamic equilibria with respect to intensive variables.

One can think here of two 'relaxation times' separated by order of magnitude. The longer relaxation time is of the order of magnitude of times taken for the macroscopic dynamical structure of the system to change. The shorter is of the order of magnitude of times taken for a single 'cell' to reach local thermodynamic equilibrium. If these two relaxation times are not well separated, then the classical non-equilibrium thermodynamical concept of local thermodynamic equilibrium loses its meaning and other approaches have to be proposed, see for instance Extended irreversible thermodynamics. For example, in the atmosphere, the speed of sound is much greater than the wind speed; this favours the idea of local thermodynamic equilibrium of matter for atmospheric heat transfer studies at altitudes below about 60 km where sound propagates, but not above 100 km, where, because of the paucity of intermolecular collisions, sound does not propagate.

Milne's 1928 Definition of Local Thermodynamic Equilibrium in Terms of Radiative Equilibrium

Milne (1928), thinking about stars, gave a definition of 'local thermodynamic equilibrium' in terms of the thermal radiation of the matter in each small local 'cell'. He defined 'local thermodynamic equilibrium' in a 'cell' by requiring that it macroscopically absorb and spontaneously emit radiation as if it were in radiative equilibrium in a cavity at the temperature of the matter of the 'cell'. Then it strictly obeys Kirchhoff's law of equality of radiative emissivity and absorptivity, with a black body source function. The key to local thermodynamic equilibrium here is that the rate of collisions of ponderable matter particles such as molecules should far exceed the rates of creation and annihilation of photons.

Entropy in Evolving Systems

It is pointed out by W.T. Grandy Jr that entropy, though it may be defined for a non-equilibrium system, is when strictly considered, only a macroscopic quantity that refers to the whole system, and is not a dynamical variable and in general does not act as a local potential that describes local physical forces. Under special circumstances, however, one can metaphorically think as if the thermal variables behaved like local physical forces. The approximation that constitutes classical irreversible thermodynamics is built on this metaphoric thinking.

This point of view shares many points in common with the concept and the use of entropy in continuum thermomechanics, which evolved completely independently of statistical mechanics and maximum-entropy principles.

Flows and Forces

The fundamental relation of classical equilibrium thermodynamics

$$dS = \frac{1}{T}dU + \frac{p}{T}dV - \sum_{i=1}^{s}\frac{\mu_i}{T}dN_i$$

expresses the change in entropy dS of a system as a function of the intensive quantities temperature T, pressure p and i^{th} chemical potential μ_i and of the differentials of the extensive quantities energy U, volume V and particle number N_i.

Following Onsager (1931,I), let us extend our considerations to thermodynamically non-equilibrium systems. As a basis, we need locally defined versions of the extensive macroscopic quantities U, V and N_i and of the intensive macroscopic quantities T, p and μ_i.

For classical non-equilibrium studies, we will consider some new locally defined intensive macroscopic variables. We can, under suitable conditions, derive these new variables by locally defining the gradients and flux densities of the basic locally defined macroscopic quantities.

Such locally defined gradients of intensive macroscopic variables are called 'thermodynamic forces'. They 'drive' flux densities, perhaps misleadingly often called 'fluxes', which are dual to the forces. These quantities are defined in the article on Onsager reciprocal relations.

Establishing the relation between such forces and flux densities is a problem in statistical mechanics. Flux densities (J_i) may be coupled. The article on Onsager reciprocal relations considers the stable near-steady thermodynamically non-equilibrium regime, which has dynamics linear in the forces and flux densities.

In stationary conditions, such forces and associated flux densities are by definition time invariant, as also are the system's locally defined entropy and rate of entropy production. Notably, according to Ilya Prigogine and others, when an open system is in conditions that allow it to reach a stable stationary thermodynamically non-equilibrium state, it organizes itself so as to minimize total entropy production defined locally.

One wants to take the analysis to the further stage of describing the behaviour of surface and volume integrals of non-stationary local quantities; these integrals are macroscopic fluxes and production rates. In general the dynamics of these integrals are not adequately described by linear equations, though in special cases they can be so described.

The Onsager Relations

Following Section III of Rayleigh (1873), Onsager (1931, I) showed that in the regime where both the flows (J_i) are small and the thermodynamic forces (F_i) vary slowly, the rate of creation of entropy (σ) is linearly related to the flows:

$$\sigma = \sum_i J_i \frac{\partial F_i}{\partial x_i}$$

and the flows are related to the gradient of the forces, parametrized by a matrix of coefficients conventionally denoted L :

$$J_i = \sum_j L_{ij} \frac{\partial F_j}{\partial x_j}$$

from which it follows that:

$$\sigma = \sum_{i,j} L_{ij} \frac{\partial F_i}{\partial x_i} \frac{\partial F_j}{\partial x_j}$$

The second law of thermodynamics requires that the matrix L be positive definite. Statistical mechanics considerations involving microscopic reversibility of dynamics imply that the matrix L is symmetric. This fact is called the *Onsager reciprocal relations*.

Speculated Extremal Principles for Non-equilibrium Processes

Until recently, prospects for useful extremal principles in this area have seemed clouded. C. Nicolis (1999) concludes that one model of atmospheric dynamics has an attractor which is not a regime of maximum or minimum dissipation; she says this seems to rule out the existence of a global organizing principle, and comments that this is to some extent disappointing; she also points to the difficulty of finding a thermodynamically consistent form of entropy production. Another top expert offers an extensive discussion of the possibilities for principles of extrema of entropy production and of dissipation of energy: Chapter 12 of Grandy (2008) is very cautious, and finds difficulty in defining the 'rate of internal entropy production' in many cases, and finds that sometimes for the prediction of the course of a process, an extremum of the quantity called the rate of dissipation of energy may be more useful than that of the rate of entropy production; this quantity appeared in Onsager's 1931 origination of this subject. Other writers have also felt that prospects for general global extremal principles are clouded. Such writers include Glansdorff and Prigogine (1971), Lebon, Jou and Casas-Vásquez (2008), and Šilhavý (1997).

A recent proposal may perhaps by-pass those clouded prospects.

Applications of Non-equilibrium Thermodynamics

Non-equilibrium thermodynamics has been successfully applied to describe biological processes such as protein folding/unfolding and transport through membranes.

Also, ideas from non-equilibrium thermodynamics and the informatic theory of entropy have been adapted to describe general economic systems.

Equilibrium State

- A system is said to be in an equilibrium state if its properties will not change without some perceivable effect in the surroundings.
- Equilibrium generally requires all properties to be uniform throughout the system.
- There are mechanical, thermal, phase, and chemical equilibria

Nature has a preferred way of directing changes. eg:

- water flows from a higher to a lower level
- Electricity flows from a higher potential to a lower one

- Heat flows from a body at higher temperature to the one at a lower temperature
- Momentum transfer occurs from a point of higher pressure to a lower one.
- Mass transfer occurs from higher concentration to a lower one

Types of Equilibrium

Between the system and surroundings, if there is no difference in

- Pressure ⇒ Mechanical equilibrium
- Potential ⇒ Electrical equilibrium
- Concentration of species ⇒ Species equilibrium
- Temperature ⇒ Thermal equilibrium

No interactions between them occur. They are said to be in equilibrium.

Thermodynamic equilibrium implies all those together.

A system in thermodynamic equilibrium does not deliver anything.

System, Surroundings and Properties

A thermodynamic system is defined as a definite quantity of matter or a region in space upon which attention is focussed in the analysis of a problem. We may want to study a quantity of matter contained with in a closed rigid walled chambers, or we may want to consider something such as gas pipeline through which the matter flows. The composition of the matter inside the system may be fixed or may change through chemical and nuclear reactions. A system may be arbitrarily defined. It becomes important when exchange of energy between the system and the everything else outside the system is considered. The judgement on the energetics of this exchange is very important.

Surroundings

Everything external to the system is surroundings. The system is distinguished from its surroundings by a specified boundary which may be at rest or in motion. The interactions between a system and its surroundings, which take place across the boundary, play an important role in thermodynamics. A system and its surroundings together comprise a universe.

Types of Systems

Two types of systems can be distinguished. These are referred to, respectively, as closed systems and open systems or control volumes. A closed system or a control mass refers to a fixed quantity of matter, whereas a control volume is a region in space through which mass may flow. A special type of closed system that does not interact with its surroundings is called an Isolated system .

Two types of exchange can occur between the system and its surroundings:

1. energy exchange (heat or work) and
2. exchange of matter (movement of molecules across the boundary of the system and surroundings).

Based on the types of exchange, one can define

- isolated systems: no exchange of matter and energy
- closed systems: no exchange of matter but some exchange of energy
- open systems: exchange of both matter and energy

If the boundary does not allow heat (energy) exchange to take place it is called adiabatic boundary otherwise it is diathermal boundary.

Property

To describe a system and predict its behaviour requires a knowledge of its properties and how those properties are related. Properties are macroscopic characteristics of a system such as mass, volume, energy, pressure and temperature to which numerical values can be assigned at a given time without knowledge of the past history of the system. Many other properties are considered during the course of our study.

- The value of a property of a system is independent of the process or the path followed by the system in reaching a particular state.
- The change in the value of the property depends only on the initial and the final states.

The word state refers to the condition of a system as described by its properties.

Mathematically, if P is a property of the system, then the change in the property in going from the initial state 1 to the final state 2 is given by

$$\int_1^2 dP = P_2 - P_1$$

If P = P (x, y) then,

$$dP = \left(\frac{\partial P}{\partial x}\right)_y dx + \left(\frac{\partial P}{\partial y}\right)_x dy = adx + bdy$$

where,

$$a = \left(\frac{\partial P}{\partial x}\right)_y \text{ and } b = \left(\frac{\partial P}{\partial y}\right)_x$$

If $\left(\frac{\partial a}{\partial y}\right) \left(\frac{\partial b}{\partial x}\right)$, then dP is said to be an exact differential, and P is a point function. A thermodynamic property is a point function and not a path function. Pressure, temperature, volume or molar volume are some of the quantities which satisfy these requirements.

Intensive and Extensive Properties

There are certain properties which depend on the size or extent of the system, and there are certain properties which are independent of the size or extent of the system. The properties like volume, which depend on the size of the system are called extensive properties. The properties, like temperature and pressure which are independent of the mass of the system are called intensive properties. The test for an intensive property is to observe how it is affected when a given system is combined with some fraction of exact replica of itself to create a new system differing only by size. Intensive properties are those which are unchanged by this process, whereas those properties whose values are increased or decreased in proportion to the enlargement or reduction of the system are called extensive properties.

Assume two identical systems S1 and S2 as shown in Figure. Both the systems are in identical states.

Let S3 be the combined system. Is the value of property for S3 same as that for S1 (and S2)?

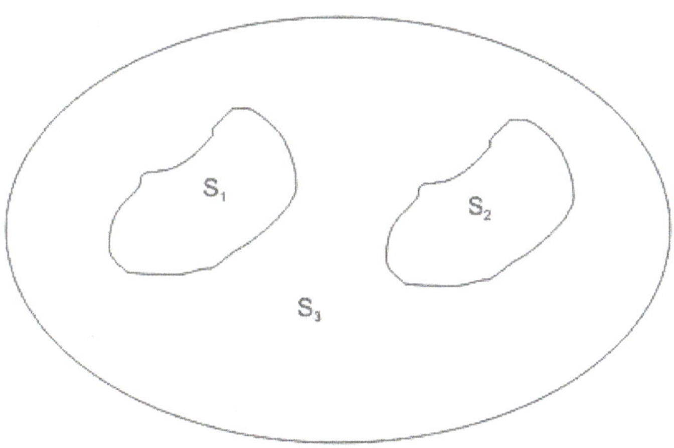

- If the answer is yes, then the property is intensive
- If the answer is no, then the property is extensive

The ratio of the extensive property to the mass is called the specific value of that property

specific volume, $v = V / m = 1/\rho$ (ρ is the density)

specific internal energy, $u = U / m$

Similarly, the molar properties are defined as the ratios of the properties to the mole number (N) of the substance

Molar volume $= \hat{v} = V / N$

Molar internal energy $= \hat{u} = U/N$

Zeroth Law of Thermodynamics

The zeroth law of thermodynamics states that if two thermodynamic systems are each in thermal equilibrium with a third, then they are in thermal equilibrium with each other. Accordingly, thermal equilibrium between systems is a transitive relation.

Two systems are said to be in the relation of thermal equilibrium if they are linked by a wall permeable only to heat and they do not change over time. As a convenience of language, systems are sometimes also said to be in a relation of thermal equilibrium if they are not linked so as to be able to transfer heat to each other, but would not do so if they were connected by a wall permeable only to heat.

The physical meaning of the law was expressed by Maxwell in the words: "All heat is of the same kind". For this reason, another statement of the law is "All diathermal walls are equivalent".

The law is important for the mathematical formulation of thermodynamics, which needs the assertion that the relation of thermal equilibrium is an equivalence relation. This information is needed for a mathematical definition of temperature that will agree with the physical existence of valid thermometers.

Zeroth Law as Equivalence Relation

A thermodynamic system is by definition in its own state of internal thermodynamic equilibrium, that is to say, there is no change in its observable state (i.e. macrostate) over time and no flows occur in it. One precise statement of the zeroth law is that the relation of thermal equilibrium is an equivalence relation on pairs of thermodynamic systems. In other words, the set of all systems each in its own state of internal thermodynamic equilibrium may be divided into subsets in which every system belongs to one and only one subset, and is in thermal equilibrium with every other member of that subset, and is not in thermal equilibrium with a member of any other subset. This means that a unique "tag" can be assigned to every system, and if the "tags" of two systems are the same, they are in thermal equilibrium with each other, and if different, they are not. This property is used to justify the use of empirical temperature as a tagging system. Empirical temperature provides further relations of thermally equilibrated systems, such as order and continuity with regard to "hotness" or "coldness", but these are not implied by the standard statement of the zeroth law.

If it is defined that a thermodynamic system is in thermal equilibrium with itself (i.e., thermal equilibrium is reflexive), then the zeroth law may be stated as follows:

If a body A, *be in thermal equilibrium with two other bodies,* B *and* C, *then* B *and* C *are in thermal equilibrium with one another.*

This statement asserts that thermal equilibrium is a left-Euclidean relation between thermodynamic systems. If we also define that every thermodynamic system is in thermal equilibrium with itself, then thermal equilibrium is also a reflexive relation. Binary relations that are both reflexive and Euclidean are equivalence relations. Thus, again implicitly assuming reflexivity, the zeroth law is therefore often expressed as a right-Euclidean statement:

If two systems are in thermal equilibrium with a third system, then they are in thermal equilibrium with each other.

One consequence of an equivalence relationship is that the equilibrium relationship is symmetric: If *A* is in thermal equilibrium with *B*, then *B* is in thermal equilibrium with *A*. Thus we may say that two systems are in thermal equilibrium with each other, or that they are in mutual equilibrium. Another consequence of equivalence is that thermal equilibrium is a transitive relationship and is occasionally expressed as such:

If A *is in thermal equilibrium with* B *and if* B *is in thermal equilibrium with* C, *then* A *is in thermal equilibrium with* C .

A reflexive, transitive relationship does not guarantee an equivalence relationship. In order for the above statement to be true, *both* reflexivity *and* symmetry must be implicitly assumed.

It is the Euclidean relationships which apply directly to thermometry. An ideal thermometer is a thermometer which does not measurably change the state of the system it is measuring. Assuming that the unchanging reading of an ideal thermometer is a valid "tagging" system for the equivalence classes of a set of equilibrated thermodynamic systems, then if a thermometer gives the same reading for two systems, those two systems are in thermal equilibrium, and if we thermally connect the two systems, there will be no subsequent change in the state of either one. If the readings are different, then thermally connecting the two systems will cause a change in the states of both systems and when the change is complete, they will both yield the same thermometer reading. The zeroth law provides no information regarding this final reading.

Foundation of Temperature

The zeroth law establishes thermal equilibrium as an equivalence relationship. An equivalence relationship on a set (such as the set of all systems each in its own state of internal thermodynamic equilibrium) divides that set into a collection of distinct subsets ("disjoint subsets") where any member of the set is a member of one and only one such subset. In the case of the zeroth law, these subsets consist of systems which are in mutual equilibrium. This partitioning allows any member of the subset to be uniquely "tagged" with a label identifying the subset to which it belongs. Although the labeling may be quite arbitrary, temperature is just such a labeling process which uses the real number system for tagging. The zeroth law justifies the use of suitable thermodynamic systems as thermometers to provide such a labeling, which yield any number of possible empirical temperature scales, and justifies the use of the second law of thermodynamics to provide an absolute, or thermodynamic temperature scale. Such temperature scales bring additional continuity and ordering (i.e., "hot" and "cold") properties to the concept of temperature.

In the space of thermodynamic parameters, zones of constant temperature form a surface, that provides a natural order of nearby surfaces. One may therefore construct a global temperature function that provides a continuous ordering of states. The dimensionality of a surface of constant temperature is one less than the number of thermodynamic parameters, thus, for an ideal gas described with three thermodynamic parameters P, V and N, it is a two-dimensional surface.

For example, if two systems of ideal gases are in equilibrium, then $P_1V_1/N_1 = P_2V_2/N_2$ where P_i is the pressure in the ith system, V_i is the volume, and N_i is the amount (in moles, or simply the number of atoms) of gas.

The surface PV/N = constant defines surfaces of equal thermodynamic temperature, and one may label defining T so that $PV/N = RT$, where R is some constant. These systems can now be used as a thermometer to calibrate other systems. Such systems are known as "ideal gas thermometers".

In a sense, focused on in the zeroth law, there is only one kind of diathermal wall or one kind of heat, as expressed by Maxwell's dictum that "All heat is of the same kind". But in another sense, heat is transferred in different ranks, as expressed by Sommerfeld's dictum "Thermodynamics investigates the conditions that govern the transformation of heat into work. It teaches us to recognize temperature as the measure of the work-value of heat. Heat of higher temperature is richer, is capable of doing more work. Work may be regarded as heat of an infinitely high temperature, as unconditionally available heat." This is why temperature is the particular variable indicated by the zeroth law's statement of equivalence.

Physical Meaning of the Usual Statement of the Zeroth Law

The present article states the zeroth law as it is often summarized in textbooks. Nevertheless, this usual statement perhaps does not explicitly convey the full physical meaning that underlies it. The underlying physical meaning was perhaps first clarified by Maxwell in his 1871 textbook.

In Carathéodory's (1909) theory, it is postulated that there exist walls "permeable only to heat", though heat is not explicitly defined in that paper. This postulate is a physical postulate of existence. It does not, however, as worded just previously, say that there is only one kind of heat. This paper of Carathéodory states as proviso 4 of its account of such walls: "Whenever each of the systems S_1 and S_2 is made to reach equilibrium with a third system S_3 under identical conditions, systems S_1 and S_2 are in mutual equilibrium". It is the function of this statement in the paper, not there labeled as the zeroth law, to provide not only for the existence of transfer of energy other than by work or transfer of matter, but further to provide that such transfer is unique in the sense that there is only one kind of such wall, and one kind of such transfer. This is signaled in the postulate of this paper of Carathéodory that precisely one non-deformation variable is needed to complete the specification of a thermodynamic state, beyond the necessary deformation variables, which are not restricted in number. It is therefore not exactly clear what Carathéodory means when in the introduction of this paper he writes "*It is possible to develop the whole theory without assuming the existence of heat, that is of a quantity that is of a different nature from the normal mechanical quantities.*"

Maxwell (1871) discusses at some length ideas which he summarizes by the words "All heat is of the same kind". Modern theorists sometimes express this idea by postulating the existence of a unique one-dimensional *hotness manifold*, into which every proper temperature scale has a monotonic mapping. This may be expressed by the statement that there is only one kind of temperature, regardless of the variety of scales in which it is expressed. Another modern expression of this idea is that "All diathermal walls are equivalent". This might also be expressed by saying that there is precisely one kind of non-mechanical, non-matter-transferring contact equilibrium between thermodynamic systems.

These ideas may be regarded as helping to clarify the physical meaning of the usual statement of the zeroth law of thermodynamics. It is the opinion of Lieb and Yngvason (1999) that the derivation from statistical mechanics of the law of entropy increase is a goal that has so far eluded the deepest thinkers. Thus the idea remains open to consideration that the existence of heat and temperature are needed as coherent primitive concepts for thermodynamics, as expressed, for example, by Maxwell and Planck. On the other hand, Planck in 1926 clarified how the second law can be stated without reference to heat or temperature, by referring to the irreversible and universal nature of friction in natural thermodynamic processes.

History

According to Arnold Sommerfeld, Ralph H. Fowler invented the title 'the zeroth law of thermodynamics' when he was discussing the 1935 text of Saha and Srivastava. They write on page 1 that "every physical quantity must be measurable in numerical terms". They presume that temperature is a physical quantity and then deduce the statement "If a body A is in temperature equilibrium with two bodies B and C, then B and C themselves will be in temperature equilibrium with each other". They then in a self-standing paragraph italicize as if to state their basic postulate: "*Any of the physical properties of A which change with the application of heat may be observed and utilised for the measurement of temperature.*" They do not themselves here use the term 'zeroth law of thermodynamics'. There are very many statements of these physical ideas in the physics literature long before this text, in very similar language. What was new here was just the label 'zeroth law of thermodynamics'. Fowler, with co-author Edward A. Guggenheim, wrote of the zeroth law as follows:

> ...we introduce the postulate: *If two assemblies are each in thermal equilibrium with a third assembly, they are in thermal equilibrium with each other.*

They then proposed that "it may be shown to follow that the condition for thermal equilibrium between several assemblies is the equality of a certain single-valued function of the thermodynamic states of the assemblies, which may be called the temperature *t*, any one of the assemblies being used as a "thermometer" reading the temperature *t* on a suitable scale. This postulate of the "*Existence of temperature*" could with advantage be known as *the zeroth law of thermodynamics*". The first sentence of this present article is a version of this statement. It is not explicitly evident in the existence statement of Fowler and Guggenheim that temperature refers to a unique attribute of a state of a system, such as is expressed in the idea of the hotness manifold. Also their statement refers explicitly to statistical mechanical assemblies, not explicitly to macroscopic thermodynamically defined systems.

If two systems (say A and B) are in thermal equilibrium with a third system (say C) separately (that is A and C are in thermal equilibrium; B and C are in thermal equilibrium) then they are in thermal equilibrium themselves (that is A and B will be in thermal equilibrium

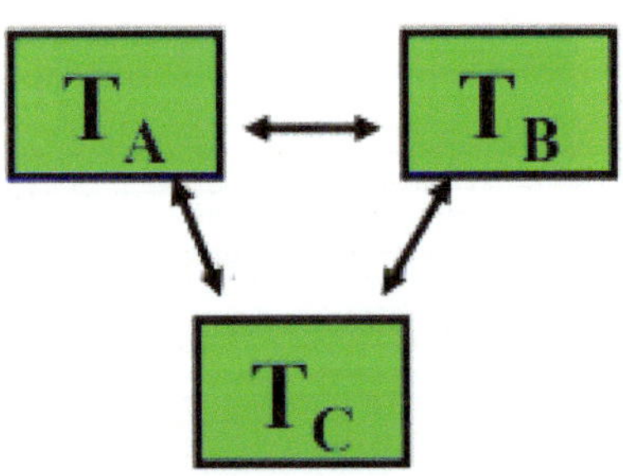

Explanation of Zeroth Law

- Let us say T_A, T_B and T_C are the temperatures of A, B and C respectively.
- A and c are in thermal equilibrium. $T_a = t_c$
- B and C are in thermal equilibrium. $T_b = t_c$

Consequence of of 'o'th law

- A and B will also be in thermal equilibrium $T_A = T_B$
- Looks very logical
- All temperature measurements are based on this LAW.

Thermal Equilibrium

Two physical systems are in thermal equilibrium if no heat flows between them when they are connected by a path permeable to heat. Thermal equilibrium obeys the zeroth law of thermodynamics. A system is said to be in thermal equilibrium with itself if the temperature within the system is spatially and temporally uniform.

Systems in thermodynamic equilibrium are always in thermal equilibrium, but the converse is not always true. If the connection between the systems allows transfer of energy as heat but does not allow transfer of matter or transfer of energy as work, the two systems may reach thermal equilibrium without reaching thermodynamic equilibrium.

Two Varieties of Thermal Equilibrium

Relation of Thermal Equilibrium between two Thermally Connected Bodies

The relation of thermal equilibrium is an instance of a contact equilibrium between two bodies, which means that it refers to transfer through a selectively permeable partition, the contact path. For the relation of thermal equilibrium, the contact path is permeable only to heat; it does not permit the passage of matter or work; it is called a diathermal connection. According to Lieb and Yngvason, the essential meaning of the relation of thermal equilibrium includes that it is reflexive and symmetric. It is not included in the essential meaning whether it is or is not transitive. After discussing the semantics of the definition, they postulate a substantial physical axiom, that they call the "zeroth law of thermodynamics", that thermal equilibrium is a transitive relation. They comment that the equivalence classes of systems so established are called isotherms.

Internal Thermal Equilibrium of an Isolated Body

Thermal equilibrium of a body in itself refers to the body when it is isolated. The background is that no heat enters or leaves it, and that it is allowed unlimited time to settle under its own intrinsic characteristics. When it is completely settled, so that macroscopic change is no longer detectable, it is in its own thermal equilibrium. It is not implied that it is necessarily in other kinds of internal equilibrium. For example, it is possible that a body might reach internal thermal equilibrium but not be in internal chemical equilibrium; glass is an example.

One may imagine an isolated system, initially not in its own state of internal thermal equilibrium. It could be subjected to a fictive thermodynamic operation of partition into two subsystems separated by nothing, no wall. One could then consider the possibility of transfers of energy as heat between the two subsystems. A long time after the fictive partition operation, the two subsystems will reach a practically stationary state, and so be in the relation of thermal equilibrium with each other. Such an adventure could be conducted in indefinitely many ways, with different fictive partitions. All of them will result in subsystems that could be shown to be in thermal equilibrium with each other, testing subsystems from different partitions. For this reason, an isolated system, initially not its own state of internal thermal equilibrium, but left for a long time, practically always will reach a final state which may be regarded as one of internal thermal equilibrium. Such a final state is one of spatial uniformity or homogeneity of temperature. The existence of such states is a basic postulate of classical thermodynamics. This postulate is sometimes, but not often, called the minus first law of thermodynamics.

Thermal Contact

Heat can flow into or out of a closed system by way of thermal conduction or of thermal radiation to or from a thermal reservoir, and when this process is effecting net transfer of heat, the system is not in thermal equilibrium. While the transfer of energy as heat continues, the system's temperature can be changing.

Bodies Prepared with Separately Uniform Temperatures, then Put into Purely Thermal Communication with each other

If bodies are prepared with separately microscopically stationary states, and are then put into purely thermal connection with each other, by conductive or radiative pathways, they will be in thermal equilibrium with each other just when the connection is followed by no change in either body. But if initially they are not in a relation of thermal equilibrium, heat will flow from the hotter to the colder, by whatever pathway, conductive or radiative, is available, and this flow will continue until thermal equilibrium is reached and then they will have the same temperature.

One form of thermal equilibrium is radiative exchange equilibrium. Two bodies, each with its own uniform temperature, in solely radiative connection, no matter how far apart, or what partially obstructive, reflective, or refractive, obstacles lie in their path of radiative exchange, not moving relative to one another, will exchange thermal radiation, in net the hotter transferring energy to the cooler, and will exchange equal and opposite amounts just when they are at the same temperature. In this situation, Kirchhoff's law of equality of radiative emissivity and absorptivity and the Helmholtz reciprocity principle are in play.

Change of Internal State of an Isolated System

If an initially isolated physical system, without internal walls that establish adiabatically isolated subsystems, is left long enough, it will usually reach a state of thermal equilibrium in itself, in which its temperature will be uniform throughout, but not necessarily a state of thermodynamic equilibrium, if there is some structural barrier that can prevent some possible processes in the system from reaching equilibrium; glass is an example. Classical thermodynamics in general considers idealized systems that have reached internal equilibrium, and idealized transfers of matter and energy between them.

An isolated physical system may be inhomogeneous, or may be composed of several subsystems separated from each other by walls. If an initially inhomogeneous physical system, without internal walls, is isolated by a thermodynamic operation, it will in general over time change its internal state. Or if it is composed of several subsystems separated from each other by walls, it may change its state after a thermodynamic operation that changes its walls. Such changes may include change of temperature or spatial distribution of temperature, by changing the state of constituent materials. A rod of iron, initially prepared to be hot at one end and cold at the other, when isolated, will change so that its temperature becomes uniform all along its length; during the process, the rod is not in thermal equilibrium until its temperature is uniform. In a system prepared as a block of ice floating in a bath of hot water, and then isolated, the ice can melt; during the melting, the system is not in thermal equilibrium; but eventually its temperature will become uniform; the block of ice will not re-form. A system prepared as a mixture of petrol vapour and air can be ignited by a spark and produce carbon dioxide and water; if this happens in an isolated system, it will increase the temperature of the system, and during the increase, the system is not in thermal equilibrium; but eventually the system will settle to a uniform temperature.

Such changes in isolated systems are irreversible in the sense that while such a change will occur spontaneously whenever the system is prepared in the same way, the reverse change will practically never occur spontaneously within the isolated system; this is a large part of the content of the second law of thermodynamics. Truly perfectly isolated systems do not occur in nature, and always are artificially prepared.

In a Gravitational Field

One may consider a system contained in a very tall adiabatically isolating vessel with rigid walls initially containing a thermally heterogeneous distribution of material, left for a long time under the influence of a steady gravitational field, along its tall dimension, due to an outside body such as the earth. It will settle to a state of uniform temperature throughout, though not of uniform pressure or density, and perhaps containing several phases. It is then in internal thermal equilibrium and even in thermodynamic equilibrium. This means that all local parts of the system are in mutual radiative exchange equilibrium. This means that the temperature of the system is spatially uniform. This is so in all cases, including those of non-uniform external force fields. For an externally imposed gravitational field, this may be proved in macroscopic thermodynamic terms, by the calculus of variations, using the method of Langrangian multipliers. Considerations of kinetic theory or statistical mechanics also support this statement.

Distinctions between Thermal and Thermodynamic Equilibria

There is an important distinction between thermal and thermodynamic equilibrium. According to Münster (1970), in states of thermodynamic equilibrium, the state variables of a system do not change at a measurable rate. Moreover, "The proviso 'at a measurable rate' implies that we can consider an equilibrium only with respect to specified processes and defined experimental conditions." Also, a state of thermodynamic equilibrium can be described by fewer macroscopic variables than any other state of a given body of matter. A single isolated body can start in a state which is not one of thermodynamic equilibrium, and can change till thermodynamic equilibrium is reached. Thermal equilibrium is a relation between two bodies or closed systems, in which transfers are allowed only of energy and take place through a partition permeable to heat, and in which the transfers have proceeded till the states of the bodies cease to change.

An explicit distinction between 'thermal equilibrium' and 'thermodynamic equilibrium' is made by C.J. Adkins. He allows that two systems might be allowed to exchange heat but be constrained from exchanging work; they will naturally exchange heat till they have equal temperatures, and reach thermal equilibrium, but in general will not be in thermodynamic equilibrium. They can reach thermodynamic equilibrium when they are allowed also to exchange work.

Another explicit distinction between 'thermal equilibrium' and 'thermodynamic equilibrium' is made by B. C. Eu. He considers two systems in thermal contact, one a thermometer, the other a system in which several irreversible processes are occurring. He considers the case in which, over the time scale of interest, it happens that both the thermometer reading and the irreversible processes are steady. Then there is thermal equilibrium without thermodynamic equilibrium. Eu proposes consequently that the zeroth law of thermodynamics can be considered to apply even when thermodynamic equilibrium is not present; also he proposes that if changes are occurring so fast that a steady temperature cannot be defined, then "it is no longer possible to describe the process by means of a thermodynamic formalism. In other words, thermodynamics has no meaning for such a process."

Temperature

- Temperature is a property of a system which determines the degree of hotness.
- Obviously, it is a relative term.

eg: A hot cup of coffee is at a higher temperature than a block of ice. On the other hand, ice is hotter than liquid hydrogen.

Thermodynamic temperature scale is under evolution. What we have now in empirical scale.

Thermodynamic Temperature

Thermodynamic temperature is the absolute measure of temperature and is one of the principal parameters of thermodynamics.

Thermodynamic temperature is defined by the third law of thermodynamics in which the theoretically lowest temperature is the null or zero point. At this point, absolute zero, the particle constituents of matter have minimal motion and can become no colder. In the quantum-mechanical description, matter at absolute zero is in its ground state, which is its state of lowest energy. Thermodynamic temperature is often also called absolute temperature, for two reasons: one, proposed by Kelvin, that it does not depend on the properties of a particular material; two that it refers to an absolute zero according to the properties of the ideal gas.

The International System of Units specifies a particular scale for thermodynamic temperature. It uses the Kelvin scale for measurement and selects the triple point of water at 273.16 K as the fundamental fixing point. Other scales have been in use historically. The Rankine scale, using the degree Fahrenheit as its unit interval, is still in use as part of the English Engineering Units in the United States in some engineering fields. ITS-90 gives a practical means of estimating the thermodynamic temperature to a very high degree of accuracy.

Roughly, the temperature of a body at rest is a measure of the mean of the energy of the translational, vibrational and rotational motions of matter's particle constituents, such as molecules, atoms, and subatomic particles. The full variety of these kinetic motions, along with potential energies of particles, and also occasionally certain other types of particle energy in equilibrium with these, make up the total internal energy of a substance. Internal energy is loosely called the heat energy or thermal energy in conditions when no work is done upon the substance by its surroundings, or by the substance upon the surroundings. Internal energy may be stored in a number of ways within a substance, each way constituting a "degree of freedom". At equilibrium, each degree of freedom will have on average the same energy: $k_B T / 2$ where k_B is the Boltzmann constant, unless that degree of freedom is in the quantum regime. The internal degrees of freedom (rotation, vibration, etc.) may be in the quantum regime at room temperature, but the translational degrees of freedom will be in the classical regime except at extremely low temperatures (fractions of kelvins) and it may be said that, for most situations, the thermodynamic temperature is specified by the average translational kinetic energy of the particles.

Overview

Temperature is a measure of the random submicroscopic motions and vibrations of the particle constituents of matter. These motions comprise the internal energy of a substance. More specifically, the thermodynamic temperature of any bulk quantity of matter is the measure of the average kinetic energy per classical (i.e., non-quantum) degree of freedom of its constituent particles. "Translational motions" are almost always in the classical regime. Translational motions are ordinary, whole-body movements in three-dimensional space in which particles move about and exchange energy in collisions. *Figure* below shows translational motion in gases; *Figure* below shows translational motion in solids. Thermodynamic temperature's null point, absolute zero, is the temperature at which the particle constituents of matter are as close as possible to complete rest; that is, they have minimal motion, retaining only quantum mechanical motion. Zero kinetic energy remains in a substance at absolute zero.

Throughout the scientific world where measurements are made in SI units, thermodynamic temperature is measured in kelvins (symbol: K). Many engineering fields in the U.S. however, measure thermodynamic temperature using the Rankine scale.

By international agreement, the unit *kelvin* and its scale are defined by two points: absolute zero, and the triple point of Vienna Standard Mean Ocean Water (water with a specified blend of hydrogen and oxygen isotopes). Absolute zero, the lowest possible temperature, is defined as being precisely 0 K *and* −273.15 °C. The triple point of water is defined as being precisely 273.16 K *and* 0.01 °C. This definition does three things:

1. It fixes the magnitude of the kelvin unit as being precisely 1 part in 273.16 parts the difference between absolute zero and the triple point of water;

2. It establishes that one kelvin has precisely the same magnitude as a one-degree increment on the Celsius scale; and

3. It establishes the difference between the two scales' null points as being precisely 273.15 kelvins (0 K = −273.15 °C and 273.16 K = 0.01 °C).

Temperatures expressed in kelvins are converted to degrees Rankine simply by multiplying by 1.8 as follows: $T_{°R} = 1.8T_K$, where T_K and $T_{°R}$ are temperatures in kelvins and degrees Rankine respectively. Temperatures expressed in degrees Rankine are converted to kelvins by *dividing* by 1.8 as follows: $T_K = {}^{T_{°R}}\!/_{1.8}$.

Practical Realization

Although the Kelvin and Celsius scales are defined using absolute zero (0 K) and the triple point of water (273.16 K and 0.01 °C), it is impractical to use this definition at temperatures that are very different from the triple point of water. ITS-90 is then designed to represent the thermodynamic temperature as closely as possible throughout its range. Many different thermometer designs are required to cover the entire range. These include helium vapor pressure thermometers, helium gas thermometers, standard platinum resistance thermometers (known as SPRTs, PRTs or Platinum RTDs) and monochromatic radiation thermometers.

For some types of thermometer the relationship between the property observed (e.g., length of a mercury column) and temperature, is close to linear, so for most purposes a linear scale is sufficient, without point-by-point calibration. For others a calibration curve or equation is required. The mercury thermometer, invented before the thermodynamic temperature was understood, originally *defined* the temperature scale; its linearity made readings correlate well with true temperature, i.e. the "mercury" temperature scale was a close fit to the true scale.

The Relationship of Temperature, Motions, Conduction, and Thermal Energy

The Nature of Kinetic Energy, Translational Motion, and Temperature

The thermodynamic temperature is a measure of the average energy of the translational, vibrational, and rotational motions of matter's particle constituents (molecules, atoms, and subatomic particles). The full variety of these kinetic motions, along with potential energies of particles, and also occasionally certain other types of particle energy in equilibrium with these, contribute the total internal energy (loosely, the thermal energy) of a substance. Thus, internal energy may be stored in a number of ways (degrees of freedom) within a substance. When the degrees of freedom are in the classical regime ("unfrozen") the temperature is very simply related to the average energy of those degrees of freedom at equilibrium. The three translational degrees of freedom are unfrozen except for the very lowest temperatures, and their kinetic energy is simply related to the thermodynamic temperature over the widest range. The heat capacity, which relates heat input and temperature change, is discussed below.

The relationship of kinetic energy, mass, and velocity is given by the formula $E_k = {}^1\!/_2 mv^2$.

Accordingly, particles with one unit of mass moving at one unit of velocity have precisely the same kinetic energy, and precisely the same temperature, as those with four times the mass but half the velocity.

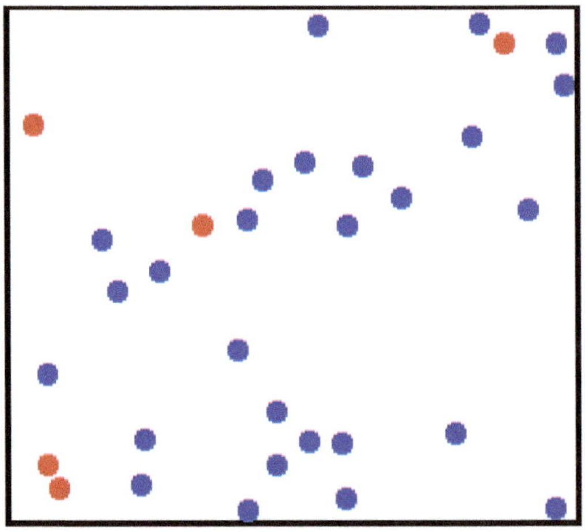

The *translational motion* of fundamental particles of nature such as atoms and molecules are directly related to temperature. Here, the size of helium atoms relative to their spacing is shown to scale under 1950 atmospheres of pressure. These room-temperature atoms have a certain average speed (slowed down here two trillion-fold). At any given instant however, a particular helium atom may be moving much faster than average while another may be nearly motionless.
Five atoms are colored red to facilitate following their motions.

Except in the quantum regime at extremely low temperatures, the thermodynamic temperature of any *bulk quantity* of a substance (a statistically significant quantity of particles) is directly proportional to the mean average kinetic energy of a specific kind of particle motion known as *translational motion*. These simple movements in the three *x*, *y*, and *z*–axis dimensions of space means the particles move in the three spatial *degrees of freedom*. The temperature derived from this translational kinetic energy is sometimes referred to as *kinetic temperature* and is equal to the thermodynamic temperature over a very wide range of temperatures. Since there are three translational degrees of freedom (e.g., motion along the x, y, and z axes), the translational kinetic energy is related to the kinetic temperature by:

$$\bar{E}=\frac{3}{2}k_B T_k$$

where:

- $\bar{E}$ is the mean kinetic energy in joules (J) and is pronounced "E bar"
- $k_B = 1.3806504(24)\times10^{-23}$ J/K is the Boltzmann constant and is pronounced "Kay sub bee"
- T_k is the kinetic temperature in kelvins (K) and is pronounced "Tee sub kay"

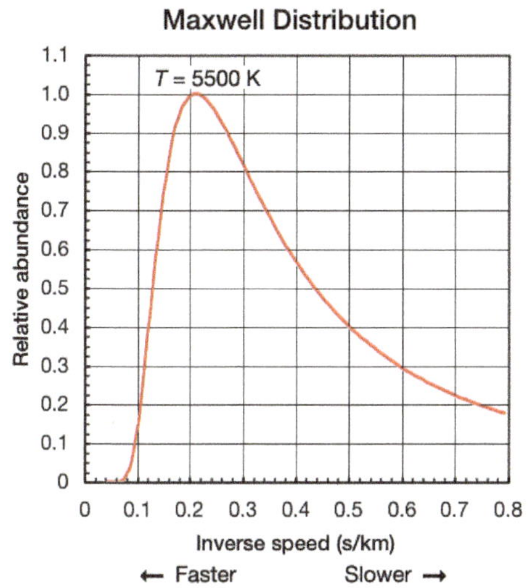

The translational motions of helium atoms occur across a range of speeds. Compare the shape of this curve to that of a Planck curve in *Figure* below.

While the Boltzmann constant is useful for finding the mean kinetic energy of a particle, it's important to note that even when a substance is isolated and in thermodynamic equilibrium (all parts are at a uniform temperature and no heat is going into or out of it), the translational motions of individual atoms and molecules occur across a wide range of speeds. At any one instant, the proportion of particles moving at a given speed within this range is determined by probability as described by the Maxwell–Boltzmann distribution. The graph shown here in *Figure* shows the speed distribution of 5500 K helium atoms. They have a *most probable* speed of 4.780 km/s. However, a certain proportion of atoms at any given instant are moving faster while others are moving relatively slowly; some are momentarily at a virtual standstill (off the x–axis to the right). This graph uses *inverse speed* for its x–axis so the shape of the curve can easily be compared to the curves in *Figure* below. In both graphs, zero on the x–axis represents infinite temperature. Additionally, the x and y–axis on both graphs are scaled proportionally.

The High Speeds of Translational Motion

Although very specialized laboratory equipment is required to directly detect translational motions, the resultant collisions by atoms or molecules with small particles suspended in a fluid produces Brownian motion that can be seen with an ordinary microscope. The translational motions of elementary particles are *very* fast and temperatures close to absolute zero are required to directly observe them. For instance, when scientists at the NIST achieved a record-setting cold temperature of 700 nK (billionths of a kelvin) in 1994, they used optical lattice laser equipment to adiabatically cool caesium atoms. They then turned off the entrapment lasers and directly measured atom velocities of 7 mm per second in order to calculate their temperature.

The Internal Motions of Molecules and Specific Heat

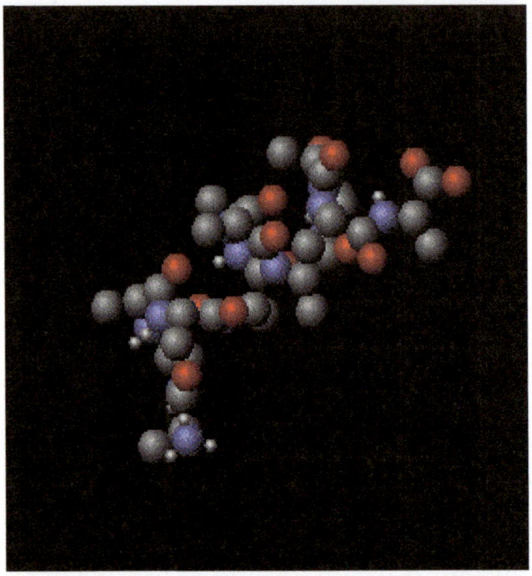

Because of their internal structure and flexibility, molecules can store kinetic energy in *internal degrees of freedom* which contribute to the heat capacity.

There are other forms of internal energy besides the kinetic energy of translational motion. As can be seen in the figure, molecules are complex objects; they are a population of atoms and thermal agitation can strain their internal chemical bonds in three different ways: via rotation, bond length, and bond angle movements. These are all types of *internal degrees of freedom*. This makes molecules distinct from *monatomic* substances (consisting of individual atoms) like the noble gases helium and argon, which have only the three translational degrees of freedom. Kinetic energy is stored in molecules' internal degrees of freedom, which gives them an *internal temperature*. Even though these motions are called *internal*, the external portions of molecules still move—rather like the jiggling of a stationary water balloon. This permits the two-way exchange of kinetic energy between internal motions and translational motions with each molecular collision. Accordingly, as energy is removed from molecules, both their kinetic temperature (the temperature derived from the kinetic energy of translational motion) and their internal temperature simultaneously diminish in equal proportions. This phenomenon is described by the equipartition theorem, which states that for any bulk quantity of a substance in equilibrium, the kinetic energy of particle motion is evenly distributed among all the active (i.e. unfrozen) degrees of freedom available to the particles. Since the internal temperature of molecules are usually equal to their kinetic temperature, the distinction is usually of interest only in the detailed study of non-local thermodynamic equilibrium (LTE) phenomena such as combustion, the sublimation of solids, and the diffusion of hot gases in a partial vacuum.

The kinetic energy stored internally in molecules causes substances to contain more internal energy at any given temperature and to absorb additional internal energy for a given temperature increase. This is because any kinetic energy that is, at a given instant, bound in internal motions is not at that same instant contributing to the molecules' translational motions. This extra thermal energy simply increases the amount of energy a substance absorbs for a given temperature rise. This property is known as a substance's specific heat capacity.

Different molecules absorb different amounts of thermal energy for each incremental increase in temperature; that is, they have different specific heat capacities. High specific heat capacity arises, in part, because certain substances' molecules possess more internal degrees of freedom than others do. For instance, nitrogen, which is a diatomic molecule, has *five* active degrees of freedom at room temperature: the three comprising translational motion plus two rotational degrees of freedom internally. Since the two internal degrees of freedom are essentially unfrozen, in accordance with the equipartition theorem, nitrogen has five-thirds the specific heat capacity per mole (a specific number of molecules) as do the monatomic gases. Another example is gasoline. Gasoline can absorb a large amount of thermal energy per mole with only a modest temperature change because each molecule comprises an average of 21 atoms and therefore has many internal degrees of freedom. Even larger, more complex molecules can have dozens of internal degrees of freedom.

The Diffusion of Thermal Energy: Entropy, Phonons, and Mobile Conduction Electrons

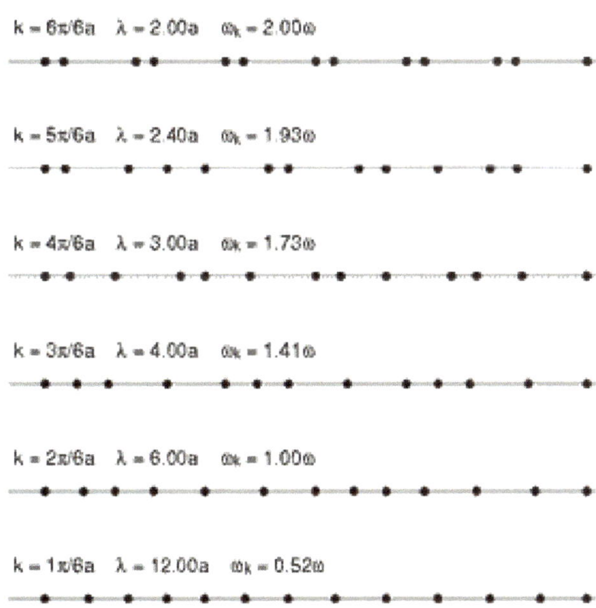

The temperature-induced translational motion of particles in solids takes the form of *phonons*. Shown here are phonons with identical amplitudes but with wavelengths ranging from 2 to 12 molecules.

Heat conduction is the diffusion of thermal energy from hot parts of a system to cold. A system can be either a single bulk entity or a plurality of discrete bulk entities. The term *bulk* in this context means a statistically significant quantity of particles (which can be a microscopic amount). Whenever thermal energy diffuses within an isolated system, temperature differences within the system decrease (and entropy increases).

One particular heat conduction mechanism occurs when translational motion, the particle motion underlying temperature, transfers momentum from particle to particle in collisions. In gases, these translational motions are of the nature shown above in *Figure*. Not only does momentum (heat) diffuse throughout the volume of the gas through serial collisions, but entire molecules or atoms can move forward into new territory, bringing their kinetic energy with them. Consequently, temperature differences equalize throughout gases very quickly—especially for light atoms or molecules; convection speeds this process even more.

Translational motion in *solids*, however, takes the form of *phonons*. Phonons are constrained, quantized wave packets that travel at a given substance's speed of sound. The manner in which phonons interact within a solid determines a variety of its properties, including its thermal conductivity. In electrically insulating solids, phonon-based heat conduction is *usually* inefficient and such solids are considered *thermal insulators* (such as glass, plastic, rubber, ceramic, and rock). This is because in solids, atoms and molecules are locked into place relative to their neighbors and are not free to roam.

Metals however, are not restricted to only phonon-based heat conduction. Thermal energy conducts through metals extraordinarily quickly because instead of direct molecule-to-molecule collisions, the vast majority of thermal energy is mediated via very light, mobile *conduction electrons*. This is why there is a near-perfect correlation between metals' thermal conductivity and their electrical conductivity. Conduction electrons imbue metals with their extraordinary conductivity because they are *delocalized* (i.e., not tied to a specific atom) and behave rather like a sort of quantum gas due to the effects of *zero-point energy*. Furthermore, electrons are relatively light with a rest mass only $^1/_{1836}$th that of a proton. This is about the same ratio as a .22 Short bullet (29 grains or 1.88 g) compared to the rifle that shoots it. As Isaac Newton wrote with his third law of motion,

Law #3: All forces occur in pairs, and these two forces are equal in magnitude and opposite in direction.

However, a bullet accelerates faster than a rifle given an equal force. Since kinetic energy increases as the square of velocity, nearly all the kinetic energy goes into the bullet, not the rifle, even though both experience the same force from the expanding propellant gases. In the same manner, because they are much less massive, thermal energy is readily borne by mobile conduction electrons. Additionally, because they're delocalized and *very* fast, kinetic thermal energy conducts extremely quickly through metals with abundant conduction electrons.

The Diffusion of Thermal Energy: Black-body Radiation

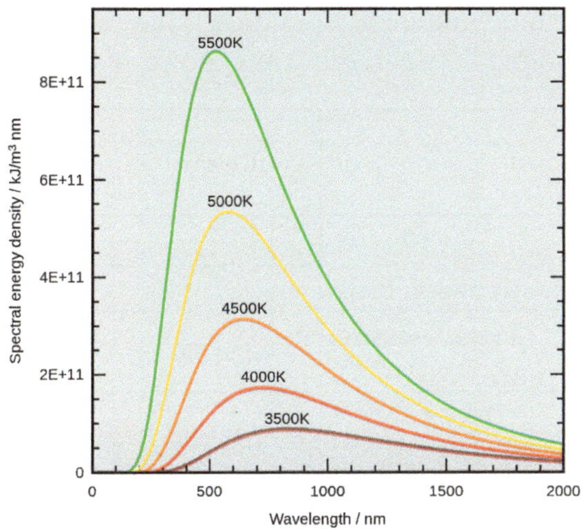

The spectrum of black-body radiation has the form of a Planck curve. A 5500 K black-body has a peak emittance wavelength of 527 nm. Compare the shape of this curve to that of a Maxwell distribution in *Figure* above.

Thermal radiation is a byproduct of the collisions arising from various vibrational motions of atoms. These collisions cause the electrons of the atoms to emit thermal photons (known as black-body radiation). Photons are emitted anytime an electric charge is accelerated (as happens when electron clouds of two atoms collide). Even *individual molecules* with internal temperatures greater than absolute zero also emit black-body radiation from their atoms. In any bulk quantity of a substance at equilibrium, black-body photons are emitted across a range of wavelengths in a spectrum that has a bell curve-like shape called a Planck curve. The top of a Planck curve (the peak emittance wavelength) is located in a particular part of the electromagnetic spectrum depending on the temperature of the black-body. Substances at extreme cryogenic temperatures emit at long radio wavelengths whereas extremely hot temperatures produce short gamma rays.

Black-body radiation diffuses thermal energy throughout a substance as the photons are absorbed by neighboring atoms, transferring momentum in the process. Black-body photons also easily escape from a substance and can be absorbed by the ambient environment; kinetic energy is lost in the process.

As established by the Stefan–Boltzmann law, the intensity of black-body radiation increases as the fourth power of absolute temperature. Thus, a black-body at 824 K (just short of glowing dull red) emits *60 times* the radiant power as it does at 296 K (room temperature). This is why one can so easily feel the radiant heat from hot objects at a distance. At higher temperatures, such as those found in an incandescent lamp, black-body radiation can be the principal mechanism by which thermal energy escapes a system.

Table of Thermodynamic Temperatures

The full range of the thermodynamic temperature scale, from absolute zero to absolute hot, and some notable points between them are shown in the table below.

	Kelvin	**Peak emittance** wavelength of black-body photons
Absolute zero (precisely by definition)	0 K	∞
Coldest measured temperature	450 pK	6,400 kilometers
One millikelvin (precisely by definition)	0.001 K	2.897 77 meters (Radio, FM band)
Cosmic Microwave Background Radiation	2.725 48(57) K	1.063 mm (peak wavelength)
Water's triple point (precisely by definition)	273.16 K	10,608.3 nm (Long wavelength I.R.)
Incandescent lamp[B]	2500 K	1160 nm (Near infrared)[C]
Sun's visible surface[C]	5778 K	501.5 nm (Green light)
Lightning bolt's channel	28,000 K	100 nm (Far Ultraviolet light)
Sun's core	16 MK	0.18 nm (X-rays)
Thermonuclear explosion (peak temperature)	350 MK	8.3×10^{-3} nm (Gamma rays)
Sandia National Labs' Z machine [D]	2 GK	1.4×10^{-3} nm (Gamma rays)
Core of a high–mass star on its last day	3 GK	1×10^{-3} nm (Gamma rays)
Merging binary neutron star system	350 GK	8×10^{-6} nm (Gamma rays)
Gamma-ray burst progenitors	1 TK	3×10^{-6} nm (Gamma rays)
Relativistic Heavy Ion Collider	1 TK	3×10^{-6} nm (Gamma rays)
CERN's proton vs. nucleus collisions	10 TK	3×10^{-7} nm (Gamma rays)
Universe 5.391×10^{-44} s after the Big Bang	1.417×10^{32} K	1.616×10^{-26} nm (Planck frequency)

A. The 2500 K value is approximate.

B. For a true blackbody (which tungsten filaments are not). Tungsten filaments' emissivity is greater at shorter wavelengths, which makes them appear whiter.

C. Effective photosphere temperature.

D. For a true blackbody (which the plasma was not). The Z machine's dominant emission originated from 40 MK electrons (soft x–ray emissions) within the plasma.

The Heat of Phase Changes

Ice and water: two phases of the same substance

The kinetic energy of particle motion is just one contributor to the total thermal energy in a substance; another is *phase transitions*, which are the potential energy of molecular bonds that can form in a substance as it cools (such as during condensing and freezing). The thermal energy required for a phase transition is called *latent heat*. This phenomenon may more easily be grasped by considering it in the reverse direction: latent heat is the energy required to *break* chemical bonds (such as during evaporation and melting). Almost everyone is familiar with the effects of phase transitions; for instance, steam at 100 °C can cause severe burns much faster than the 100 °C air from a hair dryer. This occurs because a large amount of latent heat is liberated as steam condenses into liquid water on the skin.

Even though thermal energy is liberated or absorbed during phase transitions, pure chemical elements, compounds, and eutectic alloys *exhibit no temperature change whatsoever* while they undergo them. Consider one particular type of phase transition: melting. When a solid is melting, crystal lattice chemical bonds are being broken apart; the substance is transitioning from what is known as a *more ordered state* to a *less ordered state*. In *Figure*, the melting of ice is shown within the lower left box heading from blue to green.

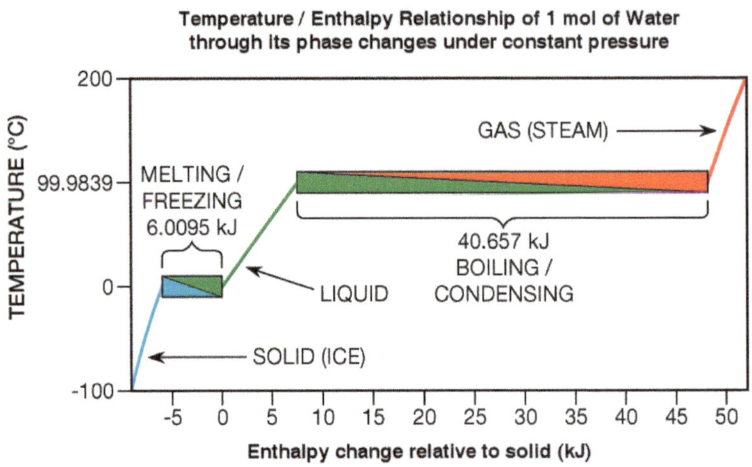

Water's temperature does not change during phase transitions as heat flows into or out of it. The total heat capacity of a mole of water in its liquid phase (the green line) is 7.5507 kJ.

At one specific thermodynamic point, the melting point (which is 0 °C across a wide pressure range in the case of water), all the atoms or molecules are, on average, at the maximum energy threshold their chemical bonds can withstand without breaking away from the lattice. Chemical bonds are all-or-nothing forces: they either hold fast, or break; there is no in-between state. Consequently, when a substance is at its melting point, every joule of added thermal energy only breaks the bonds of a specific quantity of its atoms or molecules, converting them into a liquid of precisely the same temperature; no kinetic energy is added to translational motion (which is what gives substances their temperature). The effect is rather like popcorn: at a certain temperature, additional thermal energy can't make the kernels any hotter until the transition (popping) is complete. If the process is reversed (as in the freezing of a liquid), thermal energy must be removed from a substance.

As stated above, the thermal energy required for a phase transition is called *latent heat.* In the specific cases of melting and freezing, it's called *enthalpy of fusion* or *heat of fusion.* If the molecular bonds in a crystal lattice are strong, the heat of fusion can be relatively great, typically in the range of 6 to 30 kJ per mole for water and most of the metallic elements. If the substance is one of the monatomic gases, (which have little tendency to form molecular bonds) the heat of fusion is more modest, ranging from 0.021 to 2.3 kJ per mole. Relatively speaking, phase transitions can be truly energetic events. To completely melt ice at 0 °C into water at 0 °C, one must add roughly 80 times the thermal energy as is required to increase the temperature of the same mass of liquid water by one degree Celsius. The metals' ratios are even greater, typically in the range of 400 to 1200 times. And the phase transition of boiling is much more energetic than freezing. For instance, the energy required to completely boil or vaporize water (what is known as *enthalpy of vaporization*) is roughly *540 times* that required for a one-degree increase.

Water's sizable enthalpy of vaporization is why one's skin can be burned so quickly as steam condenses on it (heading from red to green in *Figure* above). In the opposite direction, this is why one's skin feels cool as liquid water on it evaporates (a process that occurs at a sub-ambient wet-bulb temperature that is dependent on relative humidity). Water's highly energetic enthalpy of vaporization is also an important factor underlying why *solar pool covers* (floating, insulated blankets that cover swimming pools when not in use) are so effective at reducing heating costs: they prevent evaporation. For instance, the evaporation of just 20 mm of water from a 1.29-meter-deep pool chills its water 8.4 degrees Celsius (15.1 °F).

Internal Energy

The total energy of all particle motion translational and internal, including that of conduction electrons, plus the potential energy of phase changes, plus zero-point energy comprise the *internal energy* of a substance.

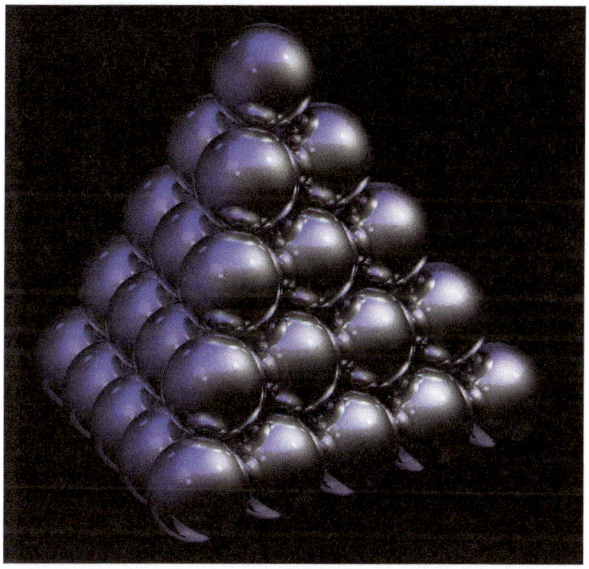

When many of the chemical elements, such as the noble gases and platinum-group metals, freeze to a solid — the most ordered state of matter — their crystal structures have a *closest-packed arrangement*. This yields the greatest possible packing density and the lowest energy state.

Internal Energy at Absolute Zero

As a substance cools, different forms of internal energy and their related effects simultaneously decrease in magnitude: the latent heat of available phase transitions is liberated as a substance changes from a less ordered state to a more ordered state; the translational motions of atoms and molecules diminish (their kinetic temperature decreases); the internal motions of molecules diminish (their internal temperature decreases); conduction electrons (if the substance is an electrical conductor) travel *somewhat* slower; and black-body radiation's peak emittance wavelength increases (the photons' energy decreases). When the particles of a substance are as close as possible to complete rest and retain only ZPE-induced quantum mechanical motion, the substance is at the temperature of absolute zero (T=0).

Note that whereas absolute zero is the point of zero thermodynamic temperature and is also the point at which the particle constituents of matter have minimal motion, absolute zero is not necessarily the point at which a substance contains zero thermal energy; one must be very precise with what one means by *internal energy*. Often, all the phase changes that *can* occur in a substance, *will* have occurred by the time it reaches absolute zero. However, this is not always the case. Notably, T=0 helium remains liquid at room pressure and must be under a pressure of at least 25 bar (2.5 MPa) to crystallize. This is because helium's heat of fusion (the energy required to melt helium ice) is so low (only 21 joules per mole) that the motion-inducing effect of zero-point energy is sufficient to prevent it from freezing at lower pressures. Only if under at least 25 bar (2.5 MPa) of pressure will this latent thermal energy be liberated as helium freezes while approaching absolute zero. A further complication is that many solids change their crystal structure to more compact arrangements at extremely high pressures (up to millions of bars, or hundreds of gigapascals). These are known as *solid-solid phase transitions* wherein latent heat is liberated as a crystal lattice changes to a more thermodynamically favorable, compact one.

The above complexities make for rather cumbersome blanket statements regarding the internal energy in T=0 substances. Regardless of pressure though, what *can* be said is that at absolute zero, all solids with a lowest-energy crystal lattice such those with a *closest-packed arrangement* contain minimal internal energy, retaining only that due to the ever-present background of zero-point energy. One can also say that for a given substance at constant pressure, absolute zero is the point of lowest *enthalpy* (a measure of work potential that takes internal energy, pressure, and volume into consideration). Lastly, it is always true to say that all T=0 substances contain zero kinetic thermal energy.

Practical Applications for Thermodynamic Temperature

Thermodynamic temperature is useful not only for scientists, it can also be useful for lay-people in many disciplines involving gases. By expressing variables in absolute terms and applying Gay–Lussac's law of temperature/pressure proportionality, solutions to everyday problems are straightforward; for instance, calculating how a temperature change affects the pressure inside an automobile tire. If the tire has a relatively cold pressure of 200 kPa-gage , then in absolute terms (relative to a vacuum), its pressure is 300 kPa-absolute. Room temperature ("cold" in tire terms) is 296 K. If the tire pressure is 20 °C hotter (20 kelvins), the solution is calculated as ${}^{316\,\mathrm{K}}/_{296\,\mathrm{K}}$ = 6.8% greater thermodynamic temperature *and* absolute pressure; that is, a pressure of 320 kPa-absolute, which is 220 kPa-gage.

Helium-4, is a superfluid at or below 2.17 kelvins,
(2.17 Celsius degrees above absolute zero)

Definition of Thermodynamic Temperature

The thermodynamic temperature is defined by the second law of thermodynamics and its consequences. The thermodynamic temperature can be shown to have special properties, and in particular can be seen to be uniquely defined (up to some constant multiplicative factor) by considering the efficiency of idealized heat engines. Thus the *ratio* T_2/T_1 of two temperaturesT_1 andT_2 is the same in all absolute scales.

Strictly speaking, the temperature of a system is well-defined only if it is at thermal equilibrium. From a microscopic viewpoint, a material is at thermal equilibrium if the quantity of heat between its individual particles cancel out. There are many possible scales of temperature, derived from a variety of observations of physical phenomena.

Loosely stated, temperature differences dictate the direction of heat between two systems such that their combined energy is maximally distributed among their lowest possible states. We call this distribution "entropy". To better understand the relationship between temperature and entropy, consider the relationship between heat, work and temperature illustrated in the Carnot heat engine. The engine converts heat into work by directing a temperature gradient between a higher temperature heat source, T_H, and a lower temperature heat sync, T_C, through a gas filled piston. The work done per cycle is equal to the difference between the heat supplied to the engine by T_H, q_H, and the heat supplied to T_C by the engine, q_C. The efficiency of the engine is the work divided by the heat put into the system or

$$\text{Efficiency} = \frac{w_{cy}}{q_H} = \frac{q_H - q_C}{q_H} = 1 - \frac{q_C}{q_H} \qquad (1)$$

where w_{cy} is the work done per cycle. Thus the efficiency depends only on q_C/q_H.

Carnot's theorem states that all reversible engines operating between the same heat reservoirs are equally efficient. Thus, any reversible heat engine operating between temperatures T_1 and T_2 must have the same efficiency, that is to say, the efficiency is the function of only temperatures

$$\frac{q_C}{q_H} = f(T_H, T_C) \qquad (2).$$

In addition, a reversible heat engine operating between temperatures T_1 and T_3 must have the same efficiency as one consisting of two cycles, one between T_1 and another (intermediate) temperature T_2, and the second between T_2 andT_3. If this were not the case, then energy (in the form of Q) will be wasted or gained, resulting in different overall efficiencies every time a cycle is split into component cycles; clearly a cycle can be composed of any number of smaller cycles.

With this understanding of Q_1, Q_2 and Q_3, we note also that mathematically,

$$f(T_1, T_3) = \frac{q_3}{q_1} = \frac{q_2 q_3}{q_1 q_2} = f(T_1, T_2) f(T_2, T_3).$$

But the first function is *NOT* a function of T_2, therefore the product of the final two functions *MUST* result in the removal of T_2 as a variable. The only way is therefore to define the function f as follows:

$$f(T_1, T_2) = \frac{g(T_2)}{g(T_1)}.$$

and

$$f(T_2, T_3) = \frac{g(T_3)}{g(T_2)}.$$

so that

$$f(T_1, T_3) = \frac{g(T_3)}{g(T_1)} = \frac{q_3}{q_1}.$$

i.e. The ratio of heat exchanged is a function of the respective temperatures at which they occur. We can choose any monotonic function for our $g(T)$; it is a matter of convenience and convention that we choose $g(T) = T$. Choosing then *one* fixed reference temperature (i.e. triple point of water), we establish the thermodynamic temperature scale.

It is to be noted that such a definition coincides with that of the ideal gas derivation; also it is this *definition* of the thermodynamic temperature that enables us to represent the Carnot efficiency in terms of T_H and T_C, and hence derive that the (complete) Carnot cycle is isentropic:

$$\frac{q_C}{q_H} = f(T_H, T_C) = \frac{T_C}{T_H}. \qquad (3).$$

Substituting this back into our first formula for efficiency yields a relationship in terms of temperature:

$$\text{Efficiency} = 1 - \frac{q_C}{q_H} = 1 - \frac{T_C}{T_H} \qquad (4).$$

Notice that for T_C=0 the efficiency is 100% and that efficiency becomes greater than 100% for T_C<0, which cases are unrealistic. Subtracting the right hand side of Equation 4 from the middle portion and rearranging gives

$$\frac{q_H}{T_H} - \frac{q_C}{T_C} = 0,$$

where the negative sign indicates heat ejected from the system. The generalization of this equation is Clausius theorem, which suggests the existence of a state function S (i.e., a function which depends only on the state of the system, not on how it reached that state) defined (up to an additive constant) by

$$dS = \frac{dq_{\text{rev}}}{T} \qquad (5),$$

where the subscript indicates heat transfer in a reversible process. The function S corresponds to the entropy of the system, mentioned previously, and the change of S around any cycle is zero (as is necessary for any state function). Equation 5 can be rearranged to get an alternative definition for temperature in terms of entropy and heat (to avoid logic loop, we should first define entropy through statistical mechanics):

$$T = \frac{dq_{\text{rev}}}{dS}.$$

For a system in which the entropy S is a function $S(E)$ of its energy E, the thermodynamic temperature T is therefore given by

$$\frac{1}{T} = \frac{dS}{dE},$$

so that the reciprocal of the thermodynamic temperature is the rate of increase of entropy with energy.

History

- Ca. 485 BC:

Parmenides

Parmenides in his treatise “On Nature” postulated the existente of *primum frigidum*, a hypothetical elementary substance source of all cooling or cold in the world.

- 1702–1703:

Guillaume Amontons

Guillaume Amontons (1663–1705) published two papers that may be used to credit him as being the first researcher to deduce the existence of a fundamental (thermodynamic) temperature scale featuring an absolute zero. He made the discovery while endeavoring to improve upon the air thermometers in use at the time. His J-tube thermometers comprised a mercury column that was supported by a fixed mass of air entrapped within the sensing portion of the thermometer. In thermodynamic terms, his thermometers relied upon the volume / temperature relationship of gas under constant pressure. His measurements of the boiling point of water and the melting point of ice showed that regardless of the mass of air trapped inside his thermometers or the weight of mercury the air was supporting, the reduction in air volume at the ice point was always the same ratio. This observation led him to posit that a sufficient reduction in temperature would reduce the air volume to zero. In fact, his calculations projected that absolute zero was equivalent to −240 °C—only 33.15 degrees short of the true value of −273.15 °C.

- 1742:

Anders Celsius

Anders Celsius (1701–1744) created a "backwards" version of the modern Celsius temperature scale. In Celsius's original scale, zero represented the boiling point of water and 100 represented the melting point of ice. In his paper *Observations of two persistent degrees on a thermometer,* he recounted his experiments showing that ice's melting point was effectively unaffected by pressure. He also determined with remarkable precision how water's boiling point varied as a function of atmospheric pressure. He proposed that zero on his temperature scale (water's boiling point) would be calibrated at the mean barometric pressure at mean sea level.

- 1744:

Carl Linnaeus

Coincident with the death of Anders Celsius, the famous botanist Carl Linnaeus (1707–1778) effectively reversed Celsius's scale upon receipt of his first thermometer featuring a scale where zero represented the melting point of ice and 100 represented water's boiling point. The custom-made *linnaeus-thermometer*, for use in his greenhouses, was made by Daniel Ekström, Sweden's leading maker of scientific instruments at the time. For the next 204 years, the scientific and thermometry communities worldwide referred to this scale as the *centigrade scale*. Temperatures on the centigrade scale were often reported simply as *degrees* or, when greater specificity was desired, *degrees centigrade*. The symbol for temperature values on this scale was °C (in several formats over the years). Because the term *centigrade* was also the French-language name for a unit of angular measurement (one-hundredth of a right angle) and had a similar connotation in other languages, the term "centesimal degree" was used when very precise, unambiguous language was required by international standards bodies such as the International Bureau of Weights and Measures (*Bureau international des poids et mesures*) (BIPM). The 9th CGPM (General Conference on Weights and Measures (*Conférence générale des poids et mesures*) and the CIPM (International Committee for Weights and Measures (*Comité international des poids et mesures*) formally adopted *degree Celsius* (symbol: °C) in 1948.

- 1777:

Johann Heinrich Lambert

In his book *Pyrometrie* (Berlin: Haude & Spener, 1779) completed four months before his death, Johann Heinrich Lambert (1728–1777), sometimes incorrectly referred to as Joseph Lambert, proposed an absolute temperature scale based on the pressure/temperature relationship of a fixed volume of gas. This is distinct from the volume/temperature relationship of gas under constant pressure that Guillaume Amontons discovered 75 years earlier. Lambert stated that absolute zero was the point where a simple straight-line extrapolation reached zero gas pressure and was equal to −270 °C.

- Circa 1787:

Jacques Alexandre César Charles

Notwithstanding the work of Guillaume Amontons 85 years earlier, Jacques Alexandre César Charles (1746–1823) is often credited with discovering, but not publishing, that the volume of a gas under constant pressure is proportional to its absolute temperature. The formula he created was $V_1/T_1 = V_2/T_2$.

- 1802:

Joseph Louis Gay-Lussac

Joseph Louis Gay-Lussac (1778–1850) published work (acknowledging the unpublished lab notes of Jacques Charles fifteen years earlier) describing how the volume of gas under constant pressure changes linearly with its absolute (thermodynamic) temperature. This behavior is called Charles's Law and is one of the gas laws. His are the first known formulas to use the number *273* for the expansion coefficient of gas relative to the melting point of ice (indicating that absolute zero was equivalent to −273 °C).

- 1848:

Lord Kelvin

William Thomson, (1824–1907) also known as Lord Kelvin, wrote in his paper, *On an Absolute Thermometric Scale,* of the need for a scale whereby *infinite cold* (absolute zero) was the scale's null point, and which used the degree Celsius for its unit increment. Like Gay-Lussac, Thomson calculated that absolute zero was equivalent to −273 °C on the air thermometers of the time. This absolute scale is known today as the Kelvin thermodynamic temperature scale. It's noteworthy that Thomson's value of *−273* was actually derived from 0.00366, which was the accepted expansion coefficient of gas per degree Celsius relative to the ice point. The inverse of −0.00366 expressed to five significant digits is −273.22 °C which is remarkably close to the true value of −273.15 °C.

- 1859:

William John Macquorn Rankine

William John Macquorn Rankine (1820–1872) proposed a thermodynamic temperature scale similar to William Thomson's but which used the degree Fahrenheit for its unit increment. This absolute scale is known today as the Rankine thermodynamic temperature scale.

- 1877–1884:

Ludwig Boltzmann

Ludwig Boltzmann (1844–1906) made major contributions to thermodynamics through an understanding of the role that particle kinetics and black body radiation played. His name is now attached to several of the formulas used today in thermodynamics.

- Circa 1930s: Gas thermometry experiments carefully calibrated to the melting point of ice and boiling point of water showed that absolute zero was equivalent to −273.15 °C.
- 1948: Resolution 3 of the 9th CGPM (Conférence Générale des Poids et Mesures, also known as the General Conference on Weights and Measures) fixed the triple point of water at precisely 0.01 °C. At this time, the triple point still had no formal definition for its equivalent kelvin value, which the resolution declared "will be fixed at a later date". The implication is that *if* the value of absolute zero measured in the 1930s was truly −273.15 °C, then the triple point of water (0.01 °C) was equivalent to 273.16 K. Additionally, both the CIPM (Comité international des poids et mesures, also known as the International Committee for Weights and Measures) and the CGPM formally adopted the name *Celsius* for the *degree Celsius* and the *Celsius temperature scale.*
- 1954: Resolution 3 of the 10th CGPM gave the Kelvin scale its modern definition by choosing the triple point of water as its second defining point and assigned it a temperature of precisely 273.16 kelvins (what was actually written 273.16 *degrees Kelvin* at the time). This, in combination with Resolution 3 of the 9th CGPM, had the effect of defining absolute zero as being precisely zero kelvins

and −273.15 °C.

- 1967/1968: Resolution 3 of the 13th CGPM renamed the unit increment of thermodynamic temperature *kelvin*, symbol K, replacing *degree absolute*, symbol °K. Further, feeling it useful to more explicitly define the magnitude of the unit increment, the 13th CGPM also decided in Resolution 4 that "The kelvin, unit of thermodynamic temperature, is the fraction 1/273.16 of the thermodynamic temperature of the triple point of water".
- 2005: The CIPM (Comité International des Poids et Mesures, also known as the International Committee for Weights and Measures) affirmed that for the purposes of delineating the temperature of the triple point of water, the definition of the Kelvin thermodynamic temperature scale would refer to water having an isotopic composition defined as being precisely equal to the nominal specification of Vienna Standard Mean Ocean Water.

References

- Eu, B.C. (2002). Generalized Thermodynamics. The Thermodynamics of Irreversible Processes and Generalized Hydrodynamics, Kluwer Academic Publishers, Dordrecht, ISBN 1-4020-0788-4
- Wildt, R. (1972). "Thermodynamics of the gray atmosphere. IV. Entropy transfer and production". Astrophysical Journal. 174: 69–77. Bibcode:1972ApJ...174...69W. doi:10.1086/151469
- Material and Energy Balances for Engineers and Environmentalists (PDF). Imperial College Press. p. 7. Retrieved May 2012
- Jou, D., Casas-Vázquez, J., Lebon, G. (1993). Extended Irreversible Thermodynamics, Springer, Berlin, ISBN 3-540-55874-8, ISBN 0-387-55874-8
- Essex, C. (1984a). "Radiation and the irreversible thermodynamics of climate". Journal of the Atmospheric Sciences. 41 (12): 1985–1991. Bibcode:1984JAtS...41.1985E. doi:10.1175/1520-0469(1984)041<1985:RATITO>2.0.CO;2
- "Hyperphysics". The Department of Physics and Astronomy of Georgia State University. Retrieved May 2012
- Lebon, G., Jou, D., Casas-Vázquez, J. (2008). Understanding Non-equilibrium Thermodynamics: Foundations, Applications, Frontiers, Springer-Verlag, Berlin, e-ISBN 978-3-540-74252-4
- Essex, C. (1984b). "Minimum entropy production in the steady state and radiative transfer". Astrophysical Journal. 285: 279–293. Bibcode:1984ApJ...285..279E. doi:10.1086/162504
- Gibbs, Willard (1993). The Scientific Papers of J. Willard Gibbs, Volume One: Thermodynamics. Ox Bow Press. ISBN 0-918024-77-3. OCLC 27974820
- Nicolis, C. (1999). "Entropy production and dynamical complexity in a low-order atmospheric model". Quarterly Journal of the Royal Meteorological Society. 125 (557): 1859–1878. Bibcode:1999QJRMS.125.1859N

2

Understanding Work Interaction Thermodynamics

Work is considered as a mode of energy transfer. Work takes place at the boundary of the system and can be observed whenever a system is undergoing a process. Another concept explored in this chapter is heat, which is the energy that flows from one body to another due to an occurrence of change in temperatures. Work and heat are one of the most significant and important topics related to thermodynamics. Different types of work include shaft work, spring work, work on elastic solids, etc. This chapter elucidates the crucial theories and principles of thermodynamics.

Work

Work is one of the basic modes of energy transfer. The work done by a system is a path function, and not a point function. Therefore, work is not a property of the system, and it cannot be said that the work is possessed by the system. It is an interaction across the boundary. What is stored in the system is energy, but not work. A decrease in energy of the system appears as work done. Therefore, work is energy in transit and it can be indentified only when the system undergoes a process.

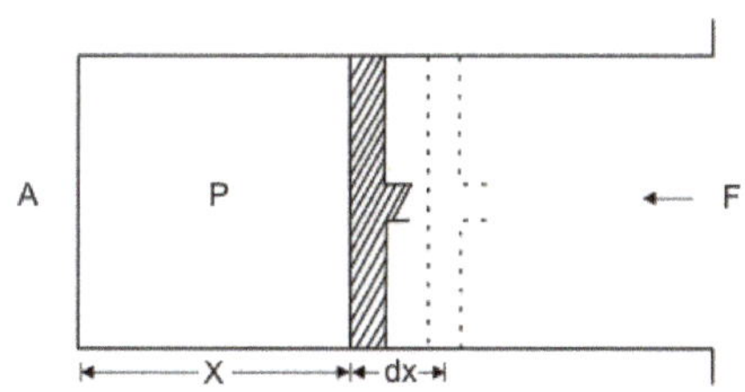

Work must be regarded only as a type of energy in transition across a well defined, zero thickness, boundary of a system. Consequently work, is never a property or any quantity contained within a system. Work is energy driven across by differences in the driving forces on either side of it. Various kinds of work are identified by the kind of driving force involved and the characteristic extensive property change which accompanied it. Work is measured quantitatively in the following way. Any driving force other than temperature, located outside the system on its external boundary, is multiplied by a transported extensive property change within the system which was transferred across the system boundary in response to this force. The result is the numerical value of the work associated with this system and driving force. In static Equilibrium, F=PA (Figure). The dX is small so that P does not change. The change in volume of the gas = AdX. The elemental work,

$$dW = FdX = PAdX = PdV$$

Thermodynamic Definition of Work

In thermodynamics, work done by a system on its surroundings during a process is defined as that interaction whose sole effect, external to the system, could be reduced as the raising of a mass through a distance against gravitational force. Let us consider the raising of mass m from an initial elevation z_1 to final elevation z_2 against gravitational force. To raise this mass, the force acting on the mass is given by F = mg . The work done on the body is $W = mg(z_2 - z_1)$

- An external agency is needed to act on the system
- It can be seen that expansion of the gas gets reduced to raising a mass against gravitational force

$$dW = F\,dX = P\,A\,dX = P\,dV$$

$$W = \int_1^2 PdV$$

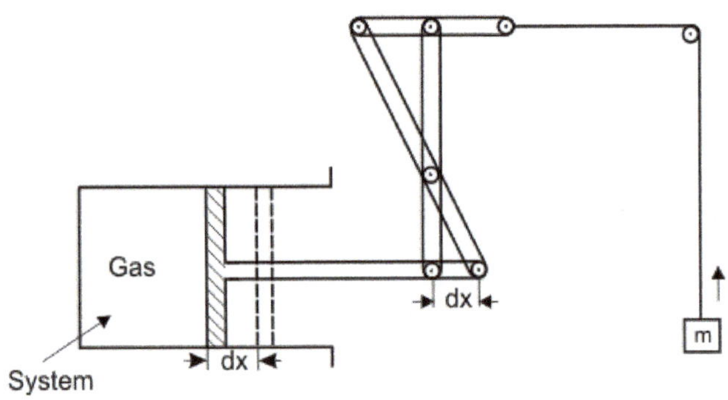

During this expansion process, the external pressure is always infinitesimally smaller than the gas pressure.

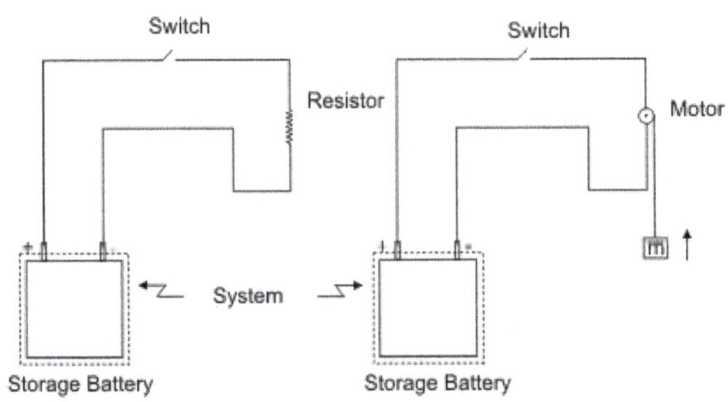

Compare two systems shown in the figure. Let the resistor be replaced by a motor drawing the same amount of current as the resistor. The motor can wind a string and thereby raise the mass which is suspended. As far as the battery is concerned, the situations are identical. So, according to thermodynamic definition of work, the interaction of a battery with a resistor is called work. By manipulating the environment, that is external to the battery (system), the effect can be reduced to raising of a mass against the gravitational force and that is the only effect on the surroundings. We can see that the thermodynamic definition of work is more general than that used in mechanics.

Situation in which $W \neq P\,dV$

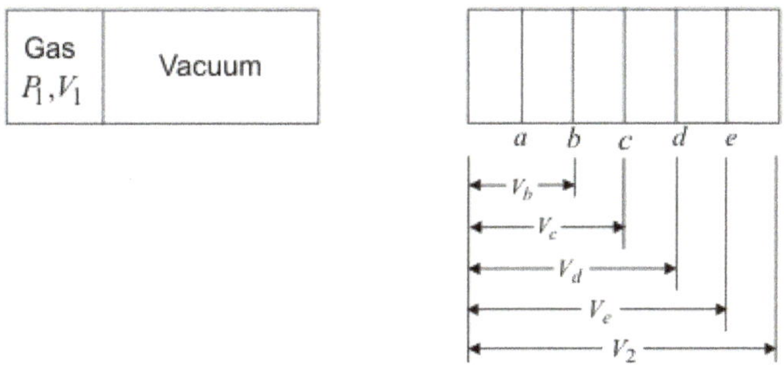

- Let the initial volume be V_1 and pressure P_1
- Let the final volume be V_2 and pressure P_2

What should be the work done in this case? Is it equal to $\int P\,dV$?

$\int P\,dV$ = area under the curve indicating the process on P-V diagram.

The expansion process may be carried out in steps as shown in figure. It is possible to draw a smooth curve passing through the points 1bcde2 . Does the area under the curve represent work done by the system? The answer is no, because the process is not reversible. The expansion of the gas is not restrained by an equal and opposing force at the moving boundary.

$$W \neq \int P\,dV$$

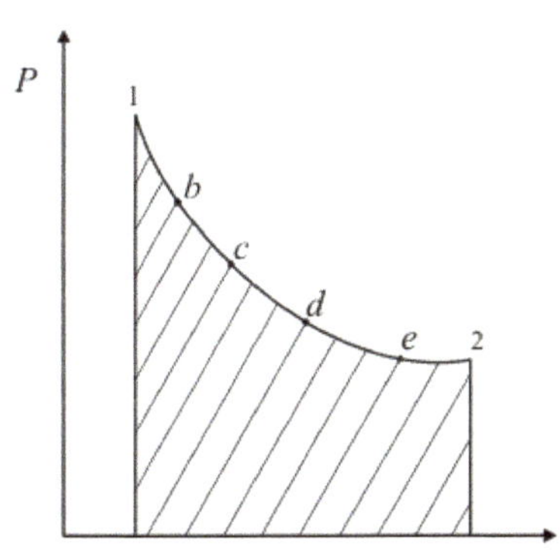

No external force has moved through any distance in this case, the work done is zero. Therefore, we observe that

- $W = \int P\, dV$ only for reversible process
- $W \neq \int P\, dV$ for an irreversible process

Another exceptional situation !

- The piston is held rigid using latches !
- $dV = 0$
- Work done on the gas is equal to the decrease in the potential energy of mass m
- A situation where $dV = 0$ and yet dW is not zero
- such work can be done in one direction only. Work is done on the system by the surroundings

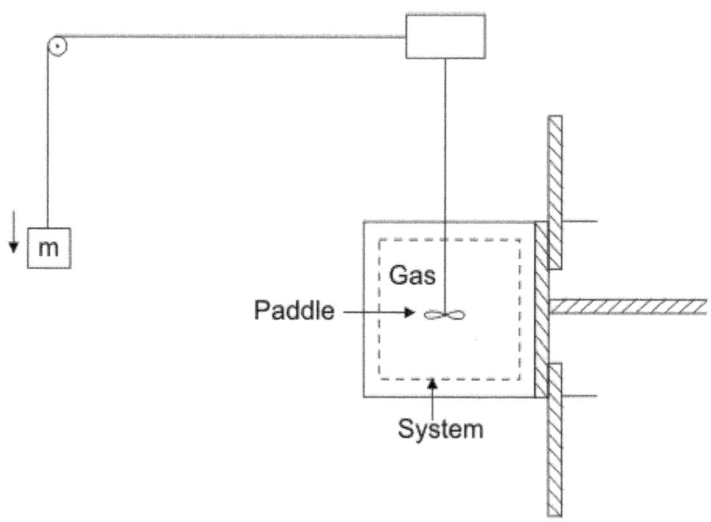

In thermodynamics, work performed by a system is the energy transferred by the system to its surroundings, that is fully accounted for solely by macroscopic forces exerted on the system by factors external to it, that is to say, factors in its surroundings. Thermodynamic work is a version of the concept of work in physics.

The external factors may be electromagnetic, gravitational, or pressure/volume or other simply mechanical constraints. Thermodynamic work is defined to be measurable solely from knowledge of such external macroscopic forces. These forces are associated with macroscopic state variables of the system that always occur in conjugate pairs, for example pressure and volume, magnetic flux density and magnetization. In the SI system of measurement, work is measured in joules (symbol: J). The rate at which work is performed is power.

History

1824

Work, i.e. "weight *lifted* through a height", was originally defined in 1824 by Sadi Carnot in his famous paper *Reflections on the Motive Power of Fire*, where he used the term *motive power*. Specifically, according to Carnot:

> *We use here motive power to express the useful effect that a motor is capable of producing. This effect can always be likened to the elevation of a weight to a certain height. It has, as we know, as a measure, the product of the weight multiplied by the height to which it is raised.*

1845

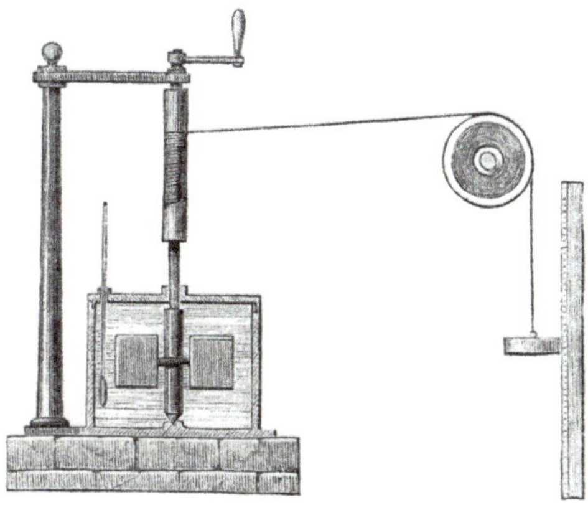

Joule's apparatus for measuring the mechanical equivalent of heat.

In 1845, the English physicist James Joule wrote a paper *On the mechanical equivalent of heat* for the British Association meeting in Cambridge. In this paper, he reported his best-known experiment, in which the *mechanical power* released through the action of a "weight *falling* through a height" was used to turn a paddle-wheel in an insulated barrel of water.

In this experiment, the friction and agitation of the paddle-wheel on the body of water caused heat to be generated which, in turn, increased the temperature of water. Both the temperature change ΔT of the water and the height of the fall Δh of the weight mg were recorded. Using these values, Joule was able to determine the mechanical equivalent of heat. Joule estimated a mechanical equivalent of heat to be 819 ft•lbf/Btu (4.41 J/cal). The modern day definitions of heat, work, temperature, and energy all have connection to this experiment.

Overview

Thermodynamic work is performed by actions such as compression, and including shaft work, stirring, and rubbing. In the simplest cases, for example, there are work of change of volume against a resisting pressure, and work without change of volume, known as isochoric work. An example of isochoric work is when an outside agency, in the surrounds of the system, drives a frictional action on the surface of the system. In this case the dissipation is usually not confined to the system, and the quantity of energy so transferred as work must be estimated through the overall change of state of the system as measured by both its mechanically and externally measurable deformation variables (such as its volume), and its corresponding non-deformation variable (such as its pressure). In a process of transfer of energy as work, the change of internal energy of the system is then defined in theory by the amount of adiabatic work that would have been necessary to reach the final from the initial state, such adiabatic work being measurable only through the externally measurable mechanical or deformation variables of the system, that provide full information about the forces exerted by the surroundings on the system during the process. In the case of some of Joule's measurements, the process was so arranged that heat produced outside the system (in the paddles) by the frictional process was practically entirely transferred into the system during the process, so that the quantity of work done by the surrounds on the system could be calculated as shaft work, an external mechanical variable.

The amount of energy transferred as work is measured through quantities defined externally to the system of interest, and thus belonging to its surroundings. In an important sign convention, work that adds to the internal energy of the system is counted as positive. Nevertheless, on the other hand, for historical reasons, an oft-encountered sign convention is to consider work done by the system on its surroundings as positive. Although all real physical processes entail some dissipation of kinetic energy, it is a matter of definition in thermodynamics that the dissipation that results from transfer of energy as work occurs only inside the system. Energy dissipated outside the system, in the process of transfer of energy, is not counted as thermodynamic work, because it is not fully accounted for by macroscopic forces exerted on the system by external factors. Thermodynamic work does not account for any energy transferred between systems as heat or though transfer of matter.

All the various mechanical and non-mechanical forms of work can be converted into each other with no fundamental limitation due to the laws of thermodynamics, so that the energy conversion efficiency can approach 100% in some cases. In particular, all forms of work can be converted into the mechanical work of lifting a weight, which was the original form of thermodynamic work considered by Carnot and Joule. Some authors have considered this equivalence to the lifting of a weight as a defining characteristic of work. In contrast, the conversion of heat into work in a heat engine can never exceed the Carnot efficiency, as a consequence of the second law of thermodynamics.

For a closed thermodynamic system, the first law of thermodynamics relates changes in the internal energy to two forms of energy transfer, as heat and as work. In theory, heat is properly defined for a process in a closed system (no transfer of matter) by the amount of adiabatic work that would be needed to effect the change occasioned by the process. In practice it is often estimated calorimetrically, through change of temperature of a known quantity of calorimetric material substance; it is of the essence of heat transfer that it is not mediated by the externally defined forces variables that define work. This distinction between work and heat is essential to thermodynamics.

Beyond the conceptual scope of thermodynamics proper, heat is transferred by the *microscopic* thermal motions of particles and their associated inter-molecular potential energies, or by radiation. There are two forms of macroscopic heat transfer by direct contact between a closed system and its surroundings: conduction, and radiation. There are several forms of dissipative transduction of energy that can occur internally within a system at a microscopic level, such as friction including bulk and shear viscosity, chemical reaction, unconstrained expansion as in Joule expansion and in diffusion, and phase change; these are not transfers of heat between systems.

Convection of internal energy is a form a transport of energy but is in general not, as sometimes mistakenly supposed (a relic of the caloric theory of heat), a form of transfer of energy as heat, because convection is not in itself a microscopic motion of microscopic particles or their intermolecular potential energies, or photons; nor is it a of transfer of energy as work. Nevertheless, if the wall between the system and its surroundings is thick and contains fluid, in the presence of a gravitational field, convective circulation within the wall can be considered as indirectly mediating transfer of energy as heat between the system and its surroundings, though they are not in direct contact.

For an open system, the first law of thermodynamics admits three forms of energy transfer, as work, as heat, and as energy associated with matter that is transferred. The latter cannot be split uniquely into heat and work components.

Formal Definition

In thermodynamics, the quantity of work done by a closed system on its surroundings is defined by factors strictly confined to the interface of the surroundings with the system and to the surroundings of the system, for example an extended gravitational field in which the system sits, that is to say, to things external to the system. There are a few especially important kinds of thermodynamic work.

A simple example of one of those important kinds is pressure-volume work. The pressure of concern is that exerted by the surroundings on the surface of the system, and the volume of interest is the negative of the increment of volume gained by the system from the surroundings. It is usually arranged that the pressure exerted by the surroundings on the surface of the system is well defined and equal to the pressure exerted by the system on the surroundings. This arrangement for transfer of energy as work can be varied in a particular way that depends on the strictly mechanical nature of pressure-volume work. The variation consists in letting the coupling between the system and surroundings be through a rigid rod that links pistons of different areas for the system and surroundings. Then for a given amount of work transferred, the exchange of volumes involves different pressures, inversely with the piston areas, for mechanical equilibrium. This cannot be done for the transfer of energy as heat because of its non-mechanical nature.

Another important kind of work is isochoric work, that is to say work that involves no eventual overall change of volume of the system between the initial and the final states of the process. Examples are friction on the surface of the system as in Rumford's experiment; shaft work such as in Joule's experiments; and slow vibrational action on the system that leaves its eventual volume unchanged, but involves friction within the system. Isochoric work for a body in its own state of internal thermodynamic equilibrium is done only by the surroundings on the body, not by the body on the surroundings, so that the sign of isochoric work with the present sign convention is always negative.

When work is done by a closed system that cannot pass heat in or out because it is adiabatically isolated, the work is referred to as being adiabatic in character. Adiabatic work can be of the pressure-volume kind or of the isochoric kind, or both.

Sign Convention

Classically, a negative value of work indicates that a positive amount of work done *by* the system leads to energy being lost from the system. This sign convention has historically been used in many physics textbooks and will be used in the present article. This sign convention also entails that a non-zero quantity of isochoric work always has a negative sign, because of the second law of thermodynamics.

According to the first law of thermodynamics for a closed system, any net increase in the internal energy U must be fully accounted for, in terms of heat δQ entering the system and the work δW done by the system:

$$dU = \delta Q - \delta W.$$

An alternate sign convention is to consider the work performed *on* the system by its surroundings as positive. This leads to a change in sign of the work, so that $dU = \delta Q - \delta W.$. This convention has historically been used in chemistry, but has been adopted several modern physics textbooks.

In the above, the letter d indicates an exact differential, expressing that internal energy U is a property of the state of the system; they depend only on the original state and the final state, and not upon the path taken. In contrast, the Greek deltas in this equation reflect the fact that the heat transfer and the work transfer are *not* properties of the final state of the system. Given only the initial state and the final state of the system, one can only say what the total change in internal energy was, not how much of the energy went out as heat, and how much as work. This can be summarized by saying that heat and work are not state functions of the system. This is in contrast to classical mechanics, where net work exerted by a particle is a state function.

Pressure-volume Work

Pressure-volume work (or *PV* work) occurs when the volume *V* of a system changes. *PV* work is often measured in units of litre-atmospheres where 1L·atm = 101.325J. However, the litre-atmosphere is not a recognised unit in the SI system of units, which measures P in Pascal (Pa), V in m^3, and PV in Joule (J), where 1 J = 1 Pa·m^3. *PV* work is an important topic in chemical thermodynamics.

For a process in a closed system, occurring slowly enough for accurate definition of the pressure on the inside of the system's wall that moves and transmits force to the surroundings, described as *quasi-static*, work is represented by the following equation between differentials:

$$\delta W = PdV$$

where

δW denotes an infinitesimal increment of work done *by* the system, transferring energy to the surroundings;

P denotes the pressure inside the system, that it exerts on the moving wall that transmits force to the surroundings. In the alternate sign convention the right hand side has a negative sign.

dV denotes the infinitesimal increment of the volume of the system.

Moreover,

$$W = \int_{V_i}^{V_f} PdV.$$

where

W denotes the work done *by* the system during the whole of the reversible process.

The first law of thermodynamics can then be expressed as

$$dU = \delta Q - PdV.$$

(In the alternate sign convention where *W* = work done *on* the system, $\delta W = -PdV$. However, $dU = \delta Q - PdV$ is unchanged.)

Path Dependence

As for all kinds of work, in general PV work is path-dependent and is therefore a thermodynamic process function. In general, the term $P\,dV$ is not an exact differential. The statement that a process is reversible and adiabatic gives important information about the process, but does not determine the path uniquely, because the path can include several slow goings backward and forward in volume, as long as there is no transfer of energy as heat. The first law of thermodynamics states $dU = \delta Q - \delta W$.. For an adiabatic process, $\delta Q = 0$ and thus the integral amount work done is equal to minus the change in internal energy. For a reversible adiabatic process, the integral amount of work done during the process depends only on the initial and final states of the process, and is the one and the same for every intermediate path.

If the process took a path other than an adiabatic path, the work would be different. This would only be possible if heat flowed into/out of the system. In a non-adiabatic process, there are indefinitely many paths between the initial and final states.

In the current mathematical notation, the differential δW is an inexact differential.

In another notation, δW is written $đW$ (with a line through the d). This notation indicates that $đW$ is not an exact one-form. The line-through is merely a flag to warn us there is actually no function (0-form) W which is the potential of $đW$. If there were, indeed, this function W, we should be able to just use Stokes Theorem to evaluate this putative function, the potential of $đW$, at the boundary of the path, that is, the initial and final points, and therefore the work would be a state function. This impossibility is consistent with the fact that it does not make sense to refer to *the work on a point* in the PV diagram; work presupposes a path.

Other Mechanical Forms of Work

There are several ways of doing work, each in some way related to a force acting through a distance. In basic mechanics, the work done by a constant force F on a body displaced a distance s in the direction of the force is given by

$$W = Fs$$

If the force is not constant, the work done is obtained by integrating the differential amount of work,

$$W = \int_1^2 F\,ds.$$

Shaft Work

Energy transmission with a rotating shaft is very common in engineering practice. Often the torque T applied to the shaft is constant which means that the force F applied is constant. For a specified constant torque, the work done during n revolutions is determined as follows: A force F acting through a moment arm r generates a torque T

$$T = Fr \rightarrow F = \frac{T}{r}$$

This force acts through a distance s, which is related to the radius r by

$$s = 2(r\pi n)$$

The shaft work is then determined from:

$$W_s = Fs = 2\pi nT$$

The power transmitted through the shaft is the shaft work done per unit time, which is expressed as

$$\dot{W}_s = 2\pi T\dot{n}$$

Spring Work

When a force is applied on a spring, and the length of the spring changes by a differential amount dx, the work done is

$$\partial w_s = Fdx$$

For linear elastic springs, the displacement x is proportional to the force applied

$$F = Kx\,,$$

where K is the spring constant and has the unit of N/m. The displacement x is measured from the undisturbed position of the spring (that is, X=0 when F=0). Substituting the two equations

$$W_s = \frac{1}{2}k\left(x_1^2 - x_2^2\right),$$

where x_1 and x_2 are the initial and the final displacement of the spring respectively, measured from the undisturbed position of the spring.

Work Done on Elastic Solid Bars

Solids are often modeled as linear springs because under the action of a force they contract or elongate, and when the force is lifted, they return to their original lengths, like a spring. This is true as long as the force is in the elastic range, that is, not large enough to cause permanent or plastic deformation. Therefore, the equations given for a linear spring can also be used for elastic solid bars. Alternately, we can determine the work associated with the expansion or contraction of an elastic solid bar by replacing the pressure P by its counterpart in solids, normal stress σ=F/A in the work expansion

$$W = \int_1^2 F\,dx.$$

$$W = \int_1^2 A\sigma\,dx.$$

where A is the cross sectional area of the bar.

Work Associated with the Stretching of Liquid Film

Consider a liquid film such as a soap film suspended on a wire frame. Some force is required to stretch this film by the movable portion of the wire frame. This force is used to overcome the microscopic forces between molecules at the liquid-air interface. These microscopic forces are perpendicular to any line in the surface and the force generated by these forces per unit length is called the surface tension σ whose unit is N/m. Therefore, the work associated with the stretching of a film is called surface tension work, and is determined from

$$W_s = \int_1^2 \sigma_s\,dA.$$

where dA=2b dx is the change in the surface area of the film. The factor 2 is due to the fact that the film has two surfaces in contact with air. The force acting on the moveable wire as a result of surface tension effects is F=2b σ, where σ is the surface tension force per unit length.

Free Energy and Exergy

The amount of useful work which may be extracted from a thermodynamic system is determined by the second law of thermodynamics. Under many practical situations this can be represented by the thermodynamic availability, or Exergy, function. Two important cases are: in thermodynamic systems where the temperature and volume are held constant, the measure of useful work attainable is the Helmholtz free energy function; and in systems where the temperature and pressure are held constant, the measure of useful work attainable is the Gibbs free energy.

Non-mechanical Forms of Work

Non-mechanical work in thermodynamics is work determined by long-range forces penetrating into the system as force fields. The action of such forces can be initiated by events in the surroundings of the system, or by thermodynamic operations on the shielding walls of the system. The long-range forces are forces in the ordinary physical sense of the word, not the so-called 'thermodynamic forces' of non-equilibrium thermodynamic terminology.

The non-mechanical work of long-range forces can have either positive or negative sign, work being done by the system on the surroundings, or *vice versa*. Work done by long-range forces can be done indefinitely slowly, so as to approach the fictive reversible quasi-static ideal, in which entropy is not created in the system by the process.

In thermodynamics, non-mechanical work is to be contrasted with mechanical work that is done by forces in immediate contact between the system and its surroundings. If the putative 'work' of a process cannot be defined as either long-range work or else as contact work, then sometimes it cannot be described by the thermodynamic formalism as work at all. Nevertheless, the thermodynamic formalism allows that energy can be transferred between an open system and its surroundings by processes for which work is not defined. An example is when the wall between the system and its surrounds is not considered as idealized and vanishingly thin, so that processes can occur within the wall, such as friction affecting the transfer of matter across the wall; in this case, the forces of transfer are neither strictly long-range nor strictly due to contact between the system and its surrounds; the transfer of energy can then be considered as by convection, and assessed in sum just as transfer of internal energy. This is conceptually different from transfer of energy as heat through a thick fluid-filled wall in the presence of a gravitational field, between a closed system and its surroundings; in this case there may convective circulation within the wall but the process may still be considered as transfer of energy as heat between the system and its surroundings; if the whole wall is moved by the application of force from the surroundings, without change of volume of the wall, so as to change the volume of the system, then it is also at the same time transferring energy as work. A chemical reaction within a system can lead to electrical long-range forces and to electric current flow, which transfer energy as work between system and surroundings, though the system's chemical reactions themselves (except for the special limiting case in which in they are driven through devices in the surroundings so as to occur along a line of thermodynamic equilibrium) are always irreversible and do not directly interact with the surroundings of the system.

Non-mechanical work contrasts with pressure-volume work. Pressure-volume work is one of the two mainly considered kinds of mechanical contact work. A force acts on the interfacing wall between system and surroundings. The force is that due to the pressure exerted on the interfacing wall by the material inside the system; that pressure is an internal state variable of the system, but is properly measured by external devices at the wall. The work is due to change of system volume by expansion or contraction of the system. If the system expands, in the present article it is said to do positive work on the surroundings. If the system contracts, in the present article it is said to do negative work on the surroundings. Pressure-volume work is a kind of contact work, because it occurs through direct material contact with the surrounding wall or matter at the boundary of the system. It is accurately described by changes in state variables of the system, such as the time courses of changes in the pressure and volume of the system. The volume of the system is classified as a "deformation variable", and is properly measured externally to the system, in the surroundings. Pressure-volume work can have either positive or negative sign. Pressure-volume work, performed slowly enough, can be made to approach the fictive reversible quasi-static ideal.

Non-mechanical work also contrasts with shaft work. Shaft work is the other of the two mainly considered kinds of mechanical contact work. It transfers energy by rotation, but it does not eventually change the shape or volume of the system. Because it does not change the volume of the system it is not measured as pressure-volume work, and it is called isochoric work. Considered solely in terms of the eventual difference between initial and final shapes and volumes of the system, shaft work does not make a change. During the process of shaft work, for example the rotation of a paddle, the shape of the system changes cyclically, but this does not make an eventual change in the shape or volume of the system. Shaft work is a kind of contact work, because it occurs through direct material contact with the surrounding matter at the boundary of the system. A system that is initially in a state of thermodynamic equilibrium cannot initiate any change in its internal energy. In particular, it cannot initiate shaft work. This explains the curious use of the phrase "inanimate material agency" by Kelvin in one of his statements of the second law of thermodynamics. Thermodynamic operations or changes in the surroundings are considered to be able to create elaborate changes such as indefinitely prolonged, varied, or ceased rotation of a driving shaft, while a system that starts in a state of thermodynamic equilibrium is inanimate and cannot spontaneously do that. Thus the sign of shaft work is always negative, work being done on the system by the surroundings. Shaft work can hardly be done indefinitely slowly; consequently it always produces entropy within the system, because it relies on friction or viscosity within the system for its transfer. The foregoing comments about shaft work apply only when one ignores that the system can store angular momentum and its related energy.

Examples of non-mechanical work modes include

- Electrical work – where the force is defined by the surroundings' *voltage* (the electrical potential) and the generalized displacement is change of spatial distribution of *electrical charge*
- Magnetic work – where the force is defined by the surroundings' *magnetic field strength* and the generalized displacement is change of total *magnetic dipole moment*
- Electrical polarization work – where the force is defined by the surroundings' *electric field strength* and the generalized displacement is change of the *polarization of the medium* (the sum of the electric dipole moments of the molecules)
- Gravitational work – where the force is defined by the surroundings' gravitational field and the generalized displacement is change of the spatial distribution of the matter within the system.

Types of work interaction

Expansion and compression work

(displacement work)

- Work of a reversible chemical cell
- Work in stretching of a liquid surface
- Work done on elastic solids
- Work of polarization and magnetization

Various Types of Work

- Displacement work (pdV work)
- Force exerted, $F = p.A$
- Work done

  $dW = F.dL = p.A\,dL = p.dV$

- If the piston moves through a finite distance say 1-2,Then work done has to be evaluated by integrating $\delta W = \int pdV$

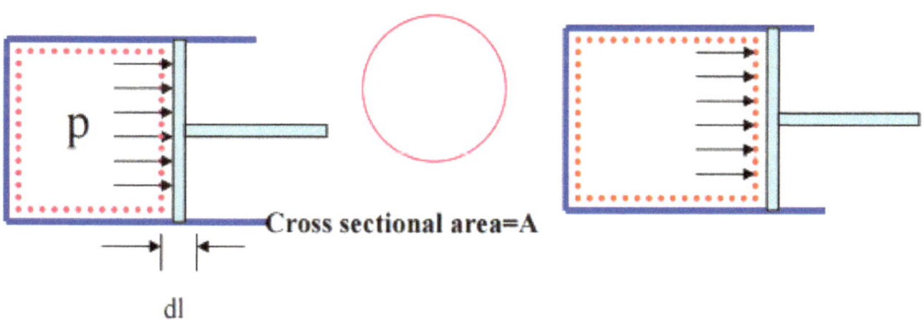

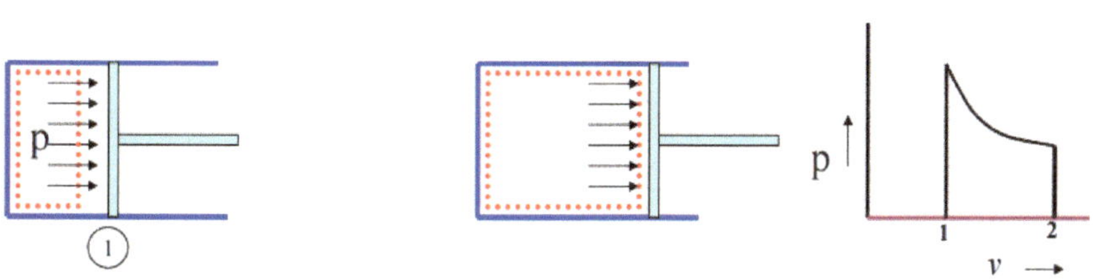

Discussion on Work Calculation

The system (shown by the dotted line) has gone through a change of state from 1 to 2.We need to know how the pressure and volume change.

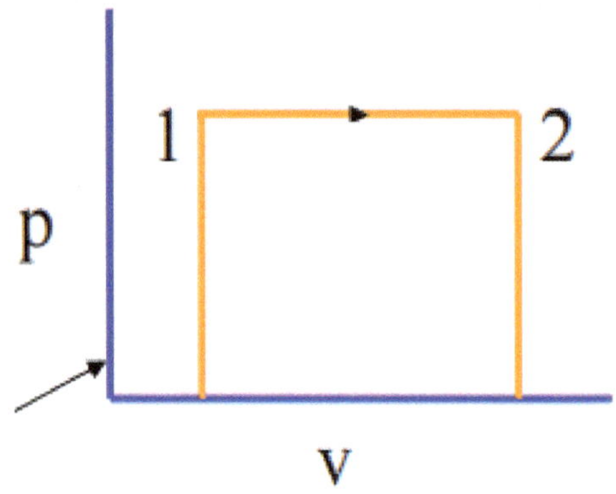

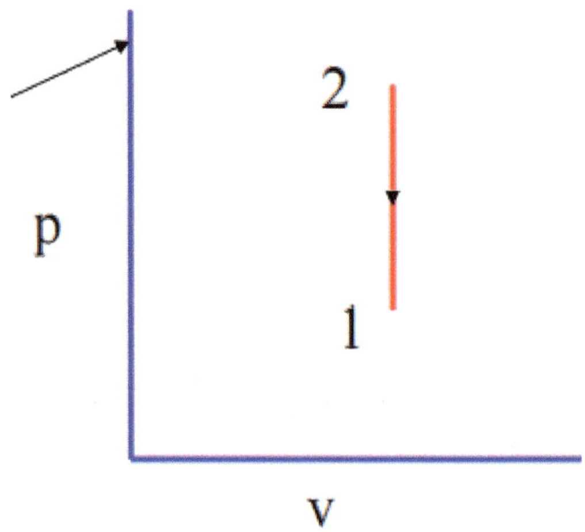

Possibilities

- Pressure might have remained constant

  or

- It might have undergone a change as per a relation p (V)

  or

- The volume might have remained constant In general the area under the process on p-V plane gives the work

Other Possible Process

- pv=constant (it will be a rectangular hyperbola)
- In general pv^n= constant

IMPORTANT: always show the states by numbers/alphabet and indicate the direction.

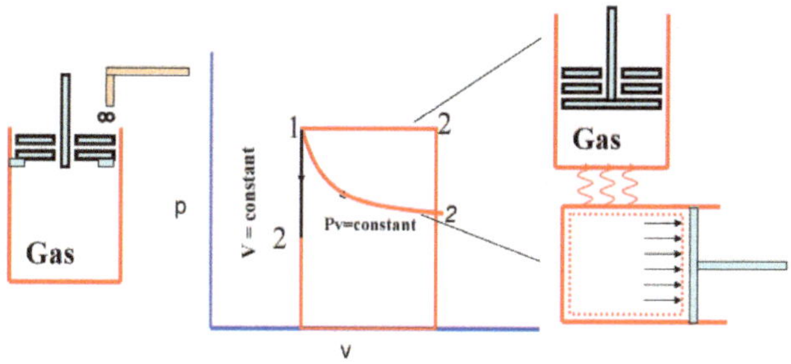

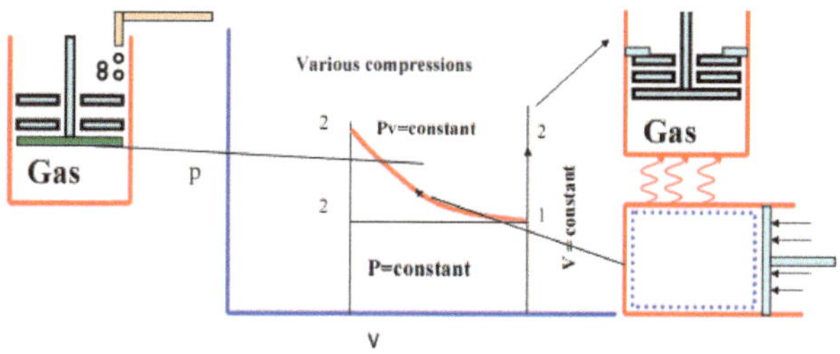

- $n = 0$ Constant pressure $(V_2 > V_1$ – **expansion**$)$
- $n = 1$ pv=constant $(p_2 < p_1 ; V_2 > V_1$ – **expansion**$)$
- $n = \infty$ Constant volume $(p_2 < p_1$ – **cooling**$)$

Others Forms Of Work

- Stretching of a wire:

Let a wire be stretched by dL due to an application of a force F Work is done on the system. Therefore $dW = -FdL$

- Electrical Energy:

Flowing in or out is always deemed to be work $dW = -EdC = -EIdt$

- Work due to stretching of a liquid film due to surface tension:

Let us say a soap film is stretched through an area dA $dW = -\sigma dA$

where σ is the surface tension.

Heat

In physics, heat is the amount of energy flowing from one body to another spontaneously due to their temperature difference, or by any means other than through work or the transfer of matter. The transfer can be by contact between the source and the destination body, as in conduction; or by radiation between remote bodies; or by way of an intermediate fluid body, as in convective circulation; or by a combination of these.

Because heat (like work) represents a quantity of energy being transferred between two bodies by certain processes, neither body "has" a definite amount of heat (much like a body in itself doesn't "have" work); in contrast, a body indeed *has* properties (state functions) such as temperature and internal energy. Thus, energy exchanged *as* heat during a given process changes the (internal) energy of each body by equal and opposite amounts. The sign of the quantity of heat can indicate the direction of the transfer, for example from system A to system B; negation indicates energy flowing in the opposite direction.

While heat flows spontaneously from hot to cold, it is possible to construct a heat pump or refrigeration system that does work to increase the difference in temperature between two systems. Conversely, a heat engine reduces an existing temperature difference to do work on another system.

Heat is a consequence of the microscopic motion of particles. When heat is transferred between two objects or systems, the energy of the object or system's particles increases. As this occurs, the arrangement between particles becomes more and more disordered. In other words, heat is related to the concept of entropy.

Historically, many energy units for measurement of heat have been used. The standards-based unit in the International System of Units (SI) is the joule (J). Heat is measured by its effect on the states of interacting bodies, for example, by the amount of ice melted or a change in temperature. The quantification of heat via the temperature change of a body is called calorimetry, and is widely used in practice. In calorimetry, sensible heat is defined with respect to a specific chosen state variable of the system, such as pressure or volume. Sensible heat causes a change of the temperature of the system while leaving the chosen state variable unchanged. Heat transfer that occurs at a constant system temperature but changes the state variable is called latent heat with respect to the variable. For infinitesimal changes, the total incremental heat transfer is then the sum of the latent and sensible heat.

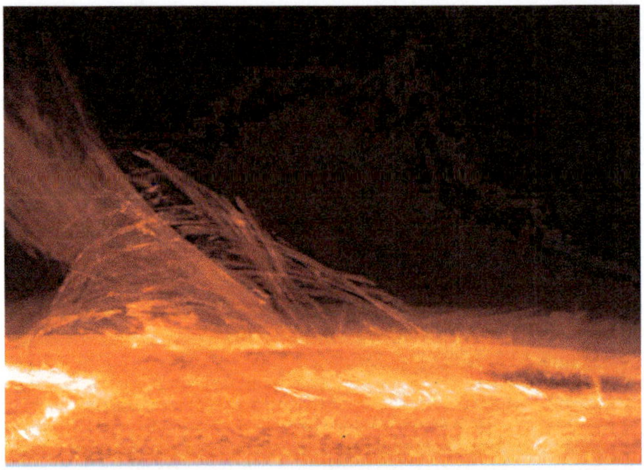

The Sun and Earth form an ongoing example of a heating process. Some of the Sun's thermal radiation strikes and heats the Earth. Compared to the Sun, Earth has a much lower temperature and so sends far less thermal radiation back to the Sun. The heat of this process can be quantified by the net amount, and direction (Sun to Earth), of energy it transferred in a given period of time.

History

Physicist James Clerk Maxwell, in his 1871 classic *Theory of Heat,* was one of many who began to build on the already established idea that heat has something to do with matter in motion. This was the same idea put forth by Benjamin Thompson in 1798, who said he was only following up on the work of many others. One of Maxwell's recommended books was *Heat as a Mode of Motion*, by John Tyndall. Maxwell outlined four stipulations for the definition of heat:

- It is *something which may be transferred from one body to another*, according to the second law of thermodynamics.
- It is a *measurable quantity*, and so can be treated mathematically.
- It *cannot be treated as a material substance,* because it may be transformed into something that is not a material substance, e.g., mechanical work.
- Heat is *one of the forms of energy.*

From empirically based ideas of heat, and from other empirical observations, the notions of internal energy and of entropy can be derived, so as to lead to the recognition of the first and second laws of thermodynamics. This was the way of the historical pioneers of thermodynamics.

Transfers of Energy as Heat between two Bodies

Referring to conduction, Partington writes: "If a hot body is brought in conducting contact with a cold body, the temperature of the hot body falls and that of the cold body rises, and it is said that a *quantity of heat* has passed from the hot body to the cold body."

Referring to radiation, Maxwell writes: "In Radiation, the hotter body loses heat, and the colder body receives heat by means of a process occurring in some intervening medium which does not itself thereby become hot."

Maxwell writes that convection as such "is not a purely thermal phenomenon". In thermodynamics, convection in general is regarded as transport of internal energy. If, however, the convection is enclosed and circulatory, then it may be regarded as an intermediary that transfers energy as heat between source and destination bodies, because it transfers only energy and not matter from the source to the destination body.

Practical Operating Devices that Harness Transfers of Energy as Heat

In accordance with the first law for closed systems, energy transferred solely as heat leaves one body and enters another, changing the internal energies of each. Transfer, between bodies, of energy as work is a complementary way of changing internal energies. Though it is not logically rigorous from the viewpoint of strict physical concepts, a common form of words that expresses this is to say that heat and work are interconvertible.

Cyclically operating engines, that use only heat and work transfers, have two thermal reservoirs, a hot and a cold one. They may be classified by the range of operating temperatures of the working body, relative to those reservoirs. In a heat engine, the working body is at all times colder than the hot reservoir and hotter than the cold reservoir. In a sense, it uses heat transfer to produce work. In a heat pump, the working body, at stages of the cycle, goes both hotter than the hot reservoir, and colder than the cold reservoir. In a sense, it uses work to produce heat transfer.

Heat Engine

In classical thermodynamics, a commonly considered model is the heat engine. It consists of four bodies: the working body, the hot reservoir, the cold reservoir, and the work reservoir. A cyclic process leaves the working body in an unchanged state, and is envisaged as being repeated indefinitely often. Work transfers between the working body and the work reservoir are envisaged as reversible, and thus only one work reservoir is needed. But two thermal reservoirs are needed, because transfer of energy as heat is irreversible. A single cycle sees energy taken by the working body from the hot reservoir and sent to the two other reservoirs, the work reservoir and the cold reservoir. The hot reservoir always and only supplies energy and the cold reservoir always and only receives energy. The second law of thermodynamics requires that no cycle can occur in which no energy is received by the cold reservoir. Heat engines achieve higher efficiency when the difference between initial and final temperature is greater.

Heat Pump or Refrigerator

Another commonly considered model is the heat pump or refrigerator. Again there are four bodies: the working body, the hot reservoir, the cold reservoir, and the work reservoir. A single cycle starts with the working body colder than the cold reservoir, and then energy is taken in as heat by the working body from the cold reservoir. Then the work reservoir does work on the working body, adding more to its internal energy, making it hotter than the hot reservoir. The hot working body passes heat to the hot reservoir, but still remains hotter than the cold reservoir. Then, by allowing it to expand without doing work on another body and without passing heat to another body, the working body is made colder than the cold reservoir. It can now accept heat transfer from the cold reservoir to start another cycle.

The device has transported energy from a colder to a hotter reservoir, but this is not regarded as by an inanimate agency; rather, it is regarded as by the harnessing of work . This is because work is supplied from the work reservoir, not just by a simple thermodynamic process, but by a cycle of thermodynamic operations and processes, which may be regarded as directed by an animate or harnessing agency. Accordingly, the cycle is still in accord with the second law of thermodynamics. The efficiency of a heat pump is best when the temperature difference between the hot and cold reservoirs is least.

Functionally, such engines are used in two ways, distinguishing a target reservoir and a resource or surrounding reservoir. A heat pump transfers heat, to the hot reservoir as the target, from the resource or surrounding reservoir. A refrigerator transfers heat, from the cold reservoir as the target, to the resource or surrounding reservoir. The target reservoir may be regarded as leaking: when the target leaks hotness to the surroundings, heat pumping is used; when the target leaks coldness to the surroundings, refrigeration is used. The engines harness work to overcome the leaks.

Macroscopic view of Quantity of Energy Transferred as Heat

According to Planck, there are three main conceptual approaches to heat. One is the microscopic or kinetic theory approach. The other two are macroscopic approaches. One is the approach through the law of conservation of energy taken as prior to thermodynamics, with a mechanical analysis of processes, for example in the work of Helmholtz. This mechanical view is taken in this article as currently customary for thermodynamic theory. The other macroscopic approach is the thermodynamic one, which admits heat as a primitive concept, which contributes, by scientific induction to knowledge of the law of conservation of energy. This view is widely taken as the practical one, quantity of heat being measured by calorimetry.

Bailyn also distinguishes the two macroscopic approaches as the mechanical and the thermodynamic. The thermodynamic view was taken by the founders of thermodynamics in the nineteenth century. It regards quantity of energy transferred as heat as a primitive concept coherent with a primitive concept of temperature, measured primarily by calorimetry. A calorimeter is a body in the surroundings of the system, with its own temperature and internal energy; when it is connected to the system by a path for heat transfer, changes in it measure heat transfer. The mechanical view was pioneered by Helmholtz and developed and used in the twentieth century, largely through the influence of Max Born. It regards quantity of heat transferred as heat as a derived concept, defined for closed systems as quantity of heat transferred by mechanisms other than work transfer, the latter being regarded as primitive for thermodynamics, defined by macroscopic mechanics. According to Born, the transfer of internal energy between open systems that accompanies transfer of matter "cannot be reduced to mechanics". It follows that there is no well-founded definition of quantities of energy transferred as heat or as work associated with transfer of matter.

Nevertheless, for the thermodynamical description of non-equilibrium processes, it is desired to consider the effect of a temperature gradient established by the surroundings across the system of interest when there is no physical barrier or wall between system and surroundings, that is to say, when they are open with respect to one another. The impossibility of a mechanical definition in terms of work for this circumstance does not alter the physical fact that a temperature gradient causes a diffusive flux of internal energy, a process that, in the thermodynamic view, might be proposed as a candidate concept for transfer of energy as heat.

In this circumstance, it may be expected that there may also be active other drivers of diffusive flux of internal energy, such as gradient of chemical potential which drives transfer of matter, and gradient of electric potential which drives electric current and iontophoresis; such effects usually interact with diffusive flux of internal energy driven by temperature gradient, and such interactions are known as cross-effects.

If cross-effects that result in diffusive transfer of internal energy were also labeled as heat transfers, they would sometimes violate the rule that pure heat transfer occurs only down a temperature gradient, never up one. They would also contradict the principle that all heat transfer is of one and the same kind, a principle founded on the idea of heat conduction between closed systems. One might to try to think narrowly of heat flux driven purely by temperature gradient as a conceptual component of diffusive internal energy flux, in the thermodynamic view, the concept resting specifically on careful calculations based on detailed knowledge of the processes and being indirectly assessed. In these circumstances, if perchance it happens that no transfer of matter is actualized, and there are no cross-effects, then the thermodynamic concept and the mechanical concept coincide, as if one were dealing with closed systems. But when there is transfer of matter, the exact laws by which temperature gradient drives diffusive flux of internal energy, rather than being exactly knowable, mostly need to be assumed, and in many cases are practically unverifiable. Consequently, when there is transfer of matter, the calculation of the pure 'heat flux' component of the diffusive flux of internal energy rests on practically unverifiable assumptions. This is a reason to think of heat as a specialized concept that relates primarily and precisely to closed systems, and applicable only in a very restricted way to open systems.

In many writings in this context, the term "heat flux" is used when what is meant is therefore more accurately called diffusive flux of internal energy; such usage of the term "heat flux" is a residue of older and now obsolete language usage that allowed that a body may have a "heat content".

Microscopic View of Heat

In the kinetic theory, heat is explained in terms of the microscopic motions and interactions of constituent particles, such as electrons, atoms, and molecules. The immediate meaning of the kinetic energy of the constituent particles is not as heat. It is as a component of internal energy. In microscopic terms, heat is a transfer quantity, and is described by a transport theory, not as steadily localized kinetic energy of particles. Heat transfer arises from temperature gradients or differences, through the diffuse exchange of microscopic kinetic and potential particle energy, by particle collisions and other interactions. An early and vague expression of this was made by Francis Bacon. Precise and detailed versions of it were developed in the nineteenth century.

In statistical mechanics, for a closed system (no transfer of matter), heat is the energy transfer associated with a disordered, microscopic action on the system, associated with jumps in occupation numbers of the energy levels of the system, without change in the values of the energy levels themselves. It is possible for macroscopic thermodynamic work to alter the occupation numbers without change in the values of the system energy levels themselves, but what distinguishes transfer as heat is that the transfer is entirely due to disordered, microscopic action, including radiative transfer. A mathematical definition can be formulated for small increments of quasi-static adiabatic work in terms of the statistical distribution of an ensemble of microstates.

Notation and Units

As a form of energy heat has the unit joule (J) in the International System of Units (SI). However, in many applied fields in engineering the British thermal unit (BTU) and the calorie are often used. The standard unit for the rate of heat transferred is the watt (W), defined as one joule per second.

The total amount of energy transferred as heat is conventionally written as Q (from Quantity) for algebraic purposes. Heat released by a system into its surroundings is by convention a negative quantity ($Q < 0$); when a system absorbs heat from its surroundings, it is positive ($Q > 0$). Heat transfer rate, or heat flow per unit time, is denoted by $\dot{Q}$. This should not be confused with a time derivative of a function of state (which can also be written with the dot notation) since heat is not a function of state. Heat flux is defined as rate of heat transfer per unit cross-sectional area, resulting in the unit *watts per square metre*.

Estimation of Quantity of Heat

Quantity of heat transferred can be measured by calorimetry, or determined through calculations based on other quantities.

Calorimetry is the empirical basis of the idea of quantity of heat transferred in a process. The transferred heat is measured by changes in a body of known properties, for example, temperature rise, change in volume or length, or phase change, such as melting of ice.

A calculation of quantity of heat transferred can rely on a hypothetical quantity of energy transferred as adiabatic work and on the first law of thermodynamics. Such calculation is the primary approach of many theoretical studies of quantity of heat transferred.

Internal Energy and Enthalpy

For a closed system (a system from which no matter can enter or exit), one version of the first law of thermodynamics states that the change in internal energy ΔU of the system is equal to the amount of heat Q supplied to the system minus the amount of work W done by system on its surroundings. The foregoing sign convention for work is used in the present article, but an alternate sign convention, followed by IUPAC, for work, is to consider the work performed on the system by its surroundings as positive. This is the convention adopted by many modern textbooks of physical chemistry, such as those by Peter Atkins and Ira Levine, but many textbooks on physics define work as work done by the system.

$$\Delta U = Q - W.$$

This formula can be re-written so as to express a definition of quantity of energy transferred as heat, based purely on the concept of adiabatic work, if it is supposed that ΔU is defined and measured solely by processes of adiabatic work:

$$Q = \Delta U + W.$$

The work done by the system includes boundary work (when the system increases its volume against an external force, such as that exerted by a piston) and other work (e.g. shaft work performed by a compressor fan), which is called isochoric work:

$$Q = \Delta U + W_{\text{boundary}} + W_{\text{isochoric}}.$$

We will neglect the "other-" or isochoric work contribution.

The internal energy, U, is a state function. In cyclical processes, such as the operation of a heat engine, state functions of the working substance return to their initial values upon completion of a cycle.

The differential, or infinitesimal increment, for the internal energy in an infinitesimal process is an exact differential $\mathrm{d}U$. The symbol for exact differentials is the lowercase letter d.

In contrast, neither of the infinitesimal increments δQ nor δW in an infinitesimal process represents the state of the system. Thus, infinitesimal increments of heat and work are inexact differentials. The lowercase Greek letter delta, δ, is the symbol for inexact differentials. The integral of any inexact differential over the time it takes for a system to leave and return to the same thermodynamic state does not necessarily equal zero.

As recounted below, in the section headed Entropy, the second law of thermodynamics observes that if heat is supplied to a system in which no irreversible processes take place and which has a well-defined temperature T, the increment of heat δQ and the temperature T form the exact differential

$$\mathrm{d}S = \frac{\delta Q}{T},$$

and that S, the entropy of the working body, is a function of state. Likewise, with a well-defined pressure, P, behind the moving boundary, the work differential, δW, and the pressure, P, combine to form the exact differential

$$dV = \frac{\delta W}{P},$$

with V the volume of the system, which is a state variable. In general, for homogeneous systems,

$$dU = TdS - PdV.$$

Associated with this differential equation is that the internal energy may be considered to be a function $U(S,V)$ of its natural variables S and V. The internal energy representation of the fundamental thermodynamic relation is written

$$U = U(S,V).$$

If V is constant

$$TdS = dU \quad (V \text{ constant})$$

and if P is constant

$$TdS = dH \quad (P \text{ constant})$$

with H the enthalpy defined by

$$H = U + PV.$$

The enthalpy may be considered to be a function $H(S,P)$ of its natural variables S and P. The enthalpy representation of the fundamental thermodynamic relation is written

$$H = H(S,P).$$

The internal energy representation and the enthalpy representation are partial Legendre transforms of one another. They contain the same physical information, written in different ways. Like the internal energy, the enthalpy stated as a function of its natural variables is a thermodynamic potential and contains all thermodynamic information about a body.

Heat Added to a Body at Constant Pressure

If a quantity Q of heat is added to a body while it does expansion work W on its surroundings, one has

$$\Delta H = \Delta U + \Delta(PV).$$

If this is constrained to happen at constant pressure with $\Delta P = 0$, the expansion work W done by the body is given by $W = P\,\Delta V$; recalling the first law of thermodynamics, one has

$$\Delta U = Q - W = Q - P\Delta V \text{ and } \Delta(PV) = P\Delta V.$$

Consequently, by substitution one has

$$\begin{aligned}\Delta H &= Q - P\Delta V + P\Delta V \\ &= Q \qquad \text{at constant pressure.}\end{aligned}$$

In this scenario, the increase in enthalpy is equal to the quantity of heat added to the system. Since many processes do take place at constant pressure, or approximately at atmospheric pressure, the enthalpy is therefore sometimes given the misleading name of 'heat content'. It is sometimes also called the heat function.

In terms of the natural variables *S* and *P* of the state function *H*, this process of change of state from state 1 to state 2 can be expressed as

$$\Delta H = \int_{S_1}^{S_2}\left(\frac{\partial H}{\partial S}\right)_P \mathrm{d}S + \int_{P_1}^{P_2}\left(\frac{\partial H}{\partial P}\right)_S \mathrm{d}P$$

$$= \int_{S_1}^{S_2}\left(\frac{\partial H}{\partial S}\right)_P \mathrm{d}S \qquad \text{at constant pressure.}$$

It is known that the temperature *T*(*S*, *P*) is identically stated by

$$\left(\frac{\partial H}{\partial S}\right)_P \equiv T(S,P).$$

Consequently,

$$\Delta H = \int_{S_1}^{S_2} T(S,P)\mathrm{d}S \qquad \text{at constant pressure.}$$

In this case, the integral specifies a quantity of heat transferred at constant pressure.

Entropy

Rudolf Clausius

In 1856, German physicist Rudolf Clausius, referring to closed systems, in which transfers of matter do not occur, defined the *second fundamental theorem* (the second law of thermodynamics) in the mechanical theory of heat (thermodynamics): "if two transformations which, without necessitating any other permanent change, can mutually replace one another, be called equivalent, then the generations of the quantity of heat Q from work at the temperature T, has the *equivalence-value*:"

$$\frac{Q}{T}.$$

In 1865, he came to define the entropy symbolized by S, such that, due to the supply of the amount of heat Q at temperature T the entropy of the system is increased by

$$\Delta S = \frac{Q}{T} \quad (1)$$

In a transfer of energy as heat without work being done, there are changes of entropy in both the surroundings which lose heat and the system which gains it. The increase, ΔS, of entropy in the system may be considered to consist of two parts, an increment, $\Delta S'$ that matches, or 'compensates', the change, $-\Delta S'$, of entropy in the surroundings, and a further increment, $\Delta S''$ that may be considered to be 'generated' or 'produced' in the system, and is said therefore to be 'uncompensated'. Thus

$$\Delta S = \Delta S' + \Delta S''.$$

This may also be written

$$\Delta S_{\text{system}} = \Delta S_{\text{compensated}} + \Delta S_{\text{uncompensated}} \text{ with } \Delta S_{\text{compensated}} = -\Delta S_{\text{surroundings}}.$$

The total change of entropy in the system and surroundings is thus

$$\Delta S_{\text{overall}} = \Delta S' + \Delta S'' - \Delta S' = \Delta S''.$$

This may also be written

$$\Delta S_{\text{overall}} = \Delta S_{\text{compensated}} + \Delta S_{\text{uncompensated}} + \Delta S_{\text{surroundings}} = \Delta S_{\text{uncompensated}}.$$

It is then said that an amount of entropy $\Delta S'$ has been transferred from the surroundings to the system. Because entropy is not a conserved quantity, this is an exception to the general way of speaking, in which an amount transferred is of a conserved quantity.

The second law of thermodynamics observes that in a natural transfer of energy as heat, in which the temperature of the system is different from that of the surroundings, it is always so that

$$\Delta S_{\text{overall}} > 0.$$

For purposes of mathematical analysis of transfers, one thinks of fictive processes that are called 'reversible', with the temperature T of the system being hardly less than that of the surroundings, and the transfer taking place at an imperceptibly slow speed.

Following the definition above in formula (1), for such a fictive 'reversible' process, a quantity of transferred heat δQ (an inexact differential) is analyzed as a quantity $T\,\mathrm{d}S$, with $\mathrm{d}S$ (an exact differential):

$$T\,\mathrm{d}S = \delta Q.$$

This equality is only valid for a fictive transfer in which there is no production of entropy, that is to say, in which there is no uncompensated entropy.

If, in contrast, the process is natural, and can really occur, with irreversibility, then there is entropy production, with $\mathrm{d}S_{\text{uncompensated}} > 0$. The quantity $T\,\mathrm{d}S_{\text{uncompensated}}$ was termed by Clausius the "uncompensated heat", though that does not accord with present-day terminology. Then one has

$$T\,\mathrm{d}S = \delta Q + T\,\mathrm{d}S_{\text{uncompensated}} > \delta Q.$$

This leads to the statement

$$T\,\mathrm{d}S \geq \delta Q \quad \text{(second law)}.$$

which is the second law of thermodynamics for closed systems.

In non-equilibrium thermodynamics that approximates by assuming the hypothesis of local thermodynamic equilibrium, there is a special notation for this. The transfer of energy as heat is assumed to take place across an infinitesimal temperature difference, so that the system element and its surroundings have near enough the same temperature T. Then one writes

$$\mathrm{d}S = \mathrm{d}S_{\mathrm{e}} + \mathrm{d}S_{\mathrm{i}},$$

where by definition

$$\delta Q = T\,\mathrm{d}S_{\mathrm{e}} \quad \text{and} \quad \mathrm{d}S_{\mathrm{i}} \equiv \mathrm{d}S_{\text{uncompensated}}.$$

The second law for a natural process asserts that

$$\mathrm{d}S_{\mathrm{i}} > 0.$$

Latent and Sensible Heat

In an 1847 lecture entitled *On Matter, Living Force, and Heat*, James Prescott Joule characterized the terms latent heat and sensible heat as components of heat each affecting distinct physical phenomena, namely the potential and kinetic energy of particles, respectively. He described latent energy as the energy possessed via a distancing of particles where attraction was over a greater distance, i.e. a form of potential energy, and the sensible heat as an energy involving the motion of particles or what was known as a *living force*. At the time of Joule kinetic energy either held 'invisibly' internally or held 'visibly' externally was known as a *living force*.

Joseph Black

Latent heat is the heat released or absorbed by a chemical substance or a thermodynamic system during a change of state that occurs without a change in temperature. Such a process may be a phase transition, such as the melting of ice or the boiling of water. The term was introduced around 1750 by Joseph Black as derived from the Latin *latere* (*to lie hidden*), characterizing its effect as not being directly measurable with a thermometer.

Sensible heat, in contrast to latent heat, is the heat transferred to a thermodynamic system that has as its sole effect a change of temperature.

Both latent heat and sensible heat transfers increase the internal energy of the system to which they are transferred.

Specific Heat

Specific heat, also called specific heat capacity, is defined as the amount of energy that has to be transferred to or from one unit of mass (kilogram) or amount of substance (mole) to change the system temperature by one degree. Specific heat is a physical property, which means that it depends on the substance under consideration and its state as specified by its properties.

The specific heats of monatomic gases (e.g., helium) are nearly constant with temperature. Diatomic gases such as hydrogen display some temperature dependence, and triatomic gases (e.g., carbon dioxide) still more.

Relation between Heat and Temperature

Before the discovery of the laws of thermodynamics, quantity of energy transferred as heat was measured by changes in the states of the participating bodies.

Most bodies, over most temperature ranges, expand on being heated. Mostly, heating a body at a constant volume increases the pressure it exerts on its constraining walls, and increases its temperature. Also mostly, heating a body at a constant pressure increases its volume, and increases its temperature.

Beyond this, most substances have three ordinarily recognized states of matter, solid, liquid, and gas, and a fourth less obviously recognized one, plasma. Many have further, more finely differentiated, states of matter, such as for example, glass, and liquid crystal. In many cases, at fixed temperature and pressure, a substance can exist in several distinct states of matter in what might be viewed as the same 'body'. For example, ice may float in a glass of water. Then the ice and the water are said to constitute two phases within the 'body'. Definite rules are known, telling how distinct phases may coexist in a 'body'. Mostly, at a fixed pressure, there is a definite temperature at which heating causes a solid to melt or evaporate, and a definite temperature at which heating causes a liquid to evaporate. In such cases, cooling has the reverse effects.

All of these, the commonest cases, fit with a rule that heating can be measured by changes of state of a body. Such cases supply what are called *thermometric bodies*, that allow the definition of empirical temperatures. Before 1848, all temperatures were defined in this way. There was thus a tight link, apparently logically determined, between heat and temperature, though they were recognized as conceptually thoroughly distinct, especially by Joseph Black in the later eighteenth century.

There are important exceptions. They break the obviously apparent link between heat and temperature. They make it clear that empirical definitions of temperature are contingent on the peculiar properties of particular thermometric substances, and are thus precluded from the title 'absolute'. For example, water contracts on being heated near 277 K. It cannot be used as a thermometric substance near that temperature. Also, over a certain temperature range, ice contracts on heating. Moreover, many substances can exist in metastable states, such as with negative pressure, that survive only transiently and in very special conditions. Such facts, sometimes called 'anomalous', are some of the reasons for the thermodynamic definition of absolute temperature.

In the early days of measurement of high temperatures, another factor was important, and used by Josiah Wedgwood in his pyrometer. The temperature reached in a process was estimated by the shrinkage of a sample of clay. The higher the temperature, the more the shrinkage. This was the only available more or less reliable method of measurement of temperatures above 1000 °C. But such shrinkage is irreversible. The clay does not expand again on cooling. That is why it could be used for the measurement. But only once. It is not a thermometric material in the usual sense of the word.

Nevertheless, the thermodynamic definition of absolute temperature does make essential use of the concept of heat, with proper circumspection.

Relation between Hotness and Temperature

According to Denbigh, the property of hotness is a concern of thermodynamics that should be defined without reference to the concept of heat. Consideration of hotness leads to the concept of empirical temperature. All physical systems are capable of heating or cooling others. This does not require that they have thermodynamic temperatures. With reference to hotness, the comparative terms hotter and colder are defined by the rule that heat flows from the hotter body to the colder.

If a physical system is inhomogeneous or very rapidly or irregularly changing, for example by turbulence, it may be impossible to characterize it by a temperature, but still there can be transfer of energy as heat between it and another system. If a system has a physical state that is regular enough, and persists long enough to allow it to reach thermal equilibrium with a specified thermometer, then it has a temperature according to that thermometer. An empirical thermometer registers degree of hotness for such a system. Such a temperature is called empirical. For example, Truesdell writes about classical thermodynamics: "At each time, the body is assigned a real number called the *temperature*. This number is a measure of how hot the body is."

Physical systems that are too turbulent to have temperatures may still differ in hotness. A physical system that passes heat to another physical system is said to be the hotter of the two. More is required for the system to have a thermodynamic temperature. Its behavior must be so regular that its empirical temperature is the same for all suitably calibrated and scaled thermometers, and then its hotness is said to lie on the one-dimensional hotness manifold. This is part of the reason why heat is defined following Carathéodory and Born, solely as occurring other than by work or transfer of matter; temperature is advisedly and deliberately not mentioned in this now widely accepted definition.

This is also the reason why the zeroth law of thermodynamics is stated explicitly. If three physical systems, *A*, *B*, and *C* are each not in their own states of internal thermodynamic equilibrium, it is possible that, with suitable physical connections being made between them, *A* can heat *B* and *B* can heat *C* and *C* can heat *A*. In non-equilibrium situations, cycles of flow are possible. It is the special and uniquely distinguishing characteristic of internal thermodynamic equilibrium that this possibility is not open to thermodynamic systems (as distinguished amongst physical systems) which are in their own states of internal thermodynamic equilibrium; this is the reason why the zeroth law of thermodynamics needs explicit statement. That is to say, the relation 'is not colder than' between general non-equilibrium physical systems is not transitive, whereas, in contrast, the relation 'has no lower a temperature than' between thermodynamic systems in their own states of internal thermodynamic equilibrium is transitive. It follows from this that the relation 'is in thermal equilibrium with' is transitive, which is one way of stating the zeroth law.

Just as temperature may undefined for a sufficiently inhomogeneous system, so also may entropy be undefined for a system not in its own state of internal thermodynamic equilibrium. For example, 'the temperature of the solar system' is not a defined quantity. Likewise, 'the entropy of the solar system' is not defined in classical thermodynamics. It has not been possible to define non-equilibrium entropy, as a simple number for a whole system, in a clearly satisfactory way.

Rigorous Definition of Quantity of Energy Transferred as Heat

It is sometimes convenient to have a rigorous definition of quantity of energy transferred as heat. Such a definition is customarily based on the work of Carathéodory (1909), referring to processes in a closed system, as follows.

The internal energy U_X of a body in an arbitrary state X can be determined by amounts of work adiabatically performed by the body on its surrounds when it starts from a reference state O. Such work is assessed through quantities defined in the surroundings of the body. It is supposed that such work can be assessed accurately, without error due to friction in the surroundings; friction in the body is not excluded by this definition. The adiabatic performance of work is defined in terms of adiabatic walls, which allow transfer of energy as work, but no other transfer, of energy or matter. In particular they do not allow the passage of energy as heat. According to this definition, work performed adiabatically is in general accompanied by friction within the thermodynamic system or body. On the other hand, according to Carathéodory (1909), there also exist non-adiabatic walls, which are postulated to be "permeable only to heat", and are called diathermal.

For the definition of quantity of energy transferred as heat, it is customarily envisaged that an arbitrary state of interest Y is reached from state O by a process with two components, one adiabatic and the other not adiabatic. For convenience one may say that the adiabatic component was the sum of work done by the body through volume change through movement of the walls while the non-adiabatic wall was temporarily rendered adiabatic, and of isochoric adiabatic work. Then the non-adiabatic component is a process of energy transfer through the wall that passes only heat, newly made accessible for the purpose of this transfer, from the surroundings to the body. The change in internal energy to reach the state Y from the state O is the difference of the two amounts of energy transferred.

Although Carathéodory himself did not state such a definition, following his work it is customary in theoretical studies to define the quantity of energy transferred as heat, Q, to the body from its surroundings, in the combined process of change to state Y from the state O, as the change in internal energy, ΔU_Y, minus the amount of work, W, done by the body on its surrounds by the adiabatic process, so that $Q = \Delta U_Y - W$.

In this definition, for the sake of conceptual rigour, the quantity of energy transferred as heat is not specified directly in terms of the non-adiabatic process. It is defined through knowledge of precisely two variables, the change of internal energy and the amount of adiabatic work done, for the combined process of change from the reference state *O* to the arbitrary state *Y*. It is important that this does not explicitly involve the amount of energy transferred in the non-adiabatic component of the combined process. It is assumed here that the amount of energy required to pass from state *O* to state *Y*, the change of internal energy, is known, independently of the combined process, by a determination through a purely adiabatic process, like that for the determination of the internal energy of state *X* above. The rigour that is prized in this definition is that there is one and only one kind of energy transfer admitted as fundamental: energy transferred as work. Energy transfer as heat is considered as a derived quantity. The uniqueness of work in this scheme is considered to guarantee rigor and purity of conception. The conceptual purity of this definition, based on the concept of energy transferred as work as an ideal notion, relies on the idea that some frictionless and otherwise non-dissipative processes of energy transfer can be realized in physical actuality. The second law of thermodynamics, on the other hand, assures us that such processes are not found in nature.

Heat, Temperature, and Thermal Equilibrium Regarded as Jointly Primitive Notions

Before the rigorous mathematical definition of heat based on Carathéodory's 1909 paper, recounted just above, historically, heat, temperature, and thermal equilibrium were presented in thermodynamics textbooks as jointly primitive notions. Carathéodory introduced his 1909 paper thus: "The proposition that the discipline of thermodynamics can be justified without recourse to any hypothesis that cannot be verified experimentally must be regarded as one of the most noteworthy results of the research in thermodynamics that was accomplished during the last century." Referring to the "point of view adopted by most authors who were active in the last fifty years", Carathéodory wrote: "There exists a physical quantity called heat that is not identical with the mechanical quantities (mass, force, pressure, etc.) and whose variations can be determined by calorimetric measurements." James Serrin introduces an account of the theory of thermodynamics thus: "In the following section, we shall use the classical notions of *heat*, *work*, and *hotness* as primitive elements, ... That heat is an appropriate and natural primitive for thermodynamics was already accepted by Carnot. Its continued validity as a primitive element of thermodynamical structure is due to the fact that it synthesizes an essential physical concept, as well as to its successful use in recent work to unify different constitutive theories." This traditional kind of presentation of the basis of thermodynamics includes ideas that may be summarized by the statement that heat transfer is purely due to spatial non-uniformity of temperature, and is by conduction and radiation, from hotter to colder bodies. It is sometimes proposed that this traditional kind of presentation necessarily rests on "circular reasoning"; against this proposal, there stands the rigorously logical mathematical development of the theory presented by Truesdell and Bharatha (1977).

This alternative approach to the definition of quantity of energy transferred as heat differs in logical structure from that of Carathéodory, recounted just above.

This alternative approach admits calorimetry as a primary or direct way to measure quantity of energy transferred as heat. It relies on temperature as one of its primitive concepts, and used in calorimetry. It is presupposed that enough processes exist physically to allow measurement of differences in internal energies. Such processes are not restricted to adiabatic transfers of energy as work. They include calorimetry, which is the commonest practical way of finding internal energy differences. The needed temperature can be either empirical or absolute thermodynamic.

In contrast, the Carathéodory way recounted just above does not use calorimetry or temperature in its primary definition of quantity of energy transferred as heat. The Carathéodory way regards calorimetry only as a secondary or indirect way of measuring quantity of energy transferred as heat. As recounted in more detail just above, the Carathéodory way regards quantity of energy transferred as heat in a process as primarily or directly defined as a residual quantity. It is calculated from the difference of the internal energies of the initial and final states of the system, and from the actual work done by the system during the process. That internal energy difference is supposed to have been measured in advance through processes of purely adiabatic transfer of energy as work, processes that take the system between the initial and final states. By the Carathéodory way it is presupposed as known from experiment that there actually physically exist enough such adiabatic processes, so that there need be no recourse to calorimetry for measurement of quantity of energy transferred as heat. This presupposition is essential but is explicitly labeled neither as a law of thermodynamics nor as an axiom of the Carathéodory way. In fact, the actual physical existence of such adiabatic processes is indeed mostly supposition, and those supposed processes have in most cases not been actually verified empirically to exist.

Heat Transfer in Engineering

A red-hot iron rod from which heat transfer to the surrounding environment will be primarily through radiation.

The discipline of heat transfer, typically considered an aspect of mechanical engineering and chemical engineering, deals with specific applied methods by which thermal energy in a system is generated, or converted, or transferred to another system. Although the definition of heat implicitly means the transfer of energy, the term *heat transfer* encompasses this traditional usage in many engineering disciplines and laymen language.

Heat transfer includes the mechanisms of heat conduction, thermal radiation, and mass transfer.

In engineering, the term *convective heat transfer* is used to describe the combined effects of conduction and fluid flow. From the thermodynamic point of view, heat flows into a fluid by diffusion to increase its energy, the fluid then transfers (advects) this increased internal energy (not heat) from one location to another, and this is then followed by a second thermal interaction which transfers heat to a second body or system, again by diffusion. This entire process is often regarded as an additional mechanism of heat transfer, although technically, "heat transfer" and thus heating and cooling occurs only on either end of such a conductive flow, but not as a result of flow. Thus, conduction can be said to "transfer" heat only as a net result of the process, but may not do so at every time within the complicated convective process.

Although distinct physical laws may describe the behavior of each of these methods, real systems often exhibit a complicated combination which are often described by a variety of mathematical methods.

Heat is energy transfer which occurs by virtue of temperature difference across the boundary. Heat like work, is energy in transit. It can be identified only at the boundary of the system. Heat is not stored in the body but energy is stored in the body. Heat like work, is not a property of the system and hence its differential is not exact. Heat and work are two different ways of transferring energy across the boundary of the system.

The displacement work is given by (figure)

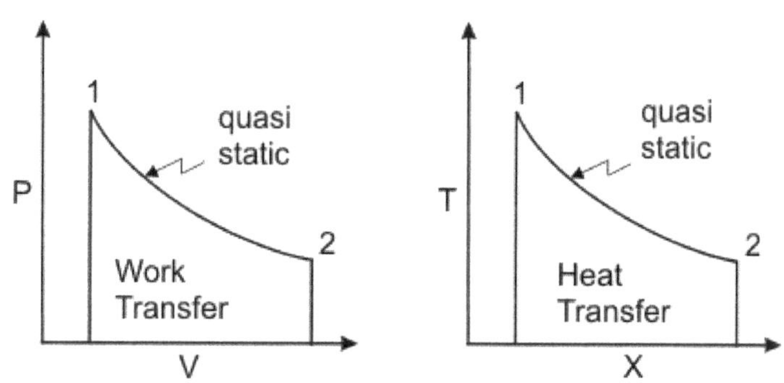

$$W_{1-2} = \int_1^2 dW = \int_1^2 PdV$$

It is valid for a quasi-static or reversible process. The work transfer is equal to the integral of the product of the intensive property P and the differential change in the extensive property dV.

Just like displacement work, the heat transfer can also be written as

$$Q_{1-2} = \int_1^2 TdX = \int_1^2 dQ$$

The quantity dQ is an inexact differential.

$$dQ = TdX$$

X is an extensive property and dX is an exact differential. The extensive property is yet to be defined. We shall see later that X is nothing but the entropy, S of a system.

It is possible to write

$$dX = \frac{1}{T}dQ$$

or,

$$\int_1^2 dX = \int_1^2 \frac{1}{T}dQ$$

Where $\frac{1}{T}$ is integrating factor.

References

- Guggenheim, E.A. (1985). Thermodynamics. An Advanced Treatment for Chemists and Physicists, seventh edition, North Holland, Amsterdam, ISBN 0444869514
- Kittel, C. Kroemer, H. (1980). Thermal Physics, second edition, W.H. Freeman, San Francisco, ISBN 0716710889.[1]
- Adkins, C.J. (1968/1983). Equilibrium Thermodynamics, (1st edition 1968), third edition 1983, Cambridge University Press, Cambridge UK, ISBN 0-521-25445-0, pp. 35–36
- Callen, H. B. (1960/1985), Thermodynamics and an Introduction to Thermostatistics, (first edition 1960), second edition 1985, John Wiley & Sons, New York, ISBN 0-471-86256-8, p. 19
- Yunus A. Cengel and Michael A. Boles,Thermodynamics: An Engineering Approach 7th Edition, McGraw-Hill, 2010,ISBN 007-352932-X
- Münster, A. (1970), Classical Thermodynamics, translated by E.S. Halberstadt, Wiley–Interscience, London, ISBN 0-471-62430-6, p. 45

3

Laws of Thermodynamics

The first law of thermodynamics follows the law of conservation of energy, which states that energy can neither be created nor destroyed. As per the first law, the increase in energy is equal to heat let out by the system. The second law states that the heat will flow from the hotter body to the colder body until the temperatures become equal. The major components of the first and second laws of thermodynamics are discussed in this section.

First Law of Thermodynamics

The first law of thermodynamics is a version of the law of conservation of energy, adapted for thermodynamic systems. The law of conservation of energy states that the total energy of an isolated system is constant; energy can be transformed from one form to another, but can be neither created nor destroyed. The first law is often formulated by stating that the change in the internal energy of a closed system is equal to the amount of heat supplied to the system, minus the amount of work done by the system on its surroundings. Equivalently, perpetual motion machines of the first kind are impossible.

History

Investigations into the nature of heat and work and their relationship began with the invention of the first engines used to extract water from mines. Improvements to such engines so as to increase their efficiency and power output came first from mechanics that worked with such machines but only slowly advanced the art. Deeper investigations that placed those on a mathematical and physics basis came later.

The first law of thermodynamics was developed empirically over about half a century. The first full statements of the law came in 1850 from Rudolf Clausius and from William Rankine; Rankine's statement is less distinct relative to Clausius'. A main aspect of the struggle was to deal with the previously proposed caloric theory of heat.

In 1840, Germain Hess stated a conservation law for the so-called 'heat of reaction' for chemical reactions. His law was later recognized as a consequence of the first law of thermodynamics, but Hess's statement was not explicitly concerned with the relation between energy exchanges by heat and work.

According to Truesdell (1980), Julius Robert von Mayer in 1841 made a statement that meant that "in a process at constant pressure, the heat used to produce expansion is universally interconvertible with work", but this is not a general statement of the first law.

Original Statements: the "Thermodynamic Approach"

The original nineteenth century statements of the first law of thermodynamics appeared in a conceptual framework in which transfer of energy as heat was taken as a primitive notion, not defined or constructed by the theoretical development of the framework, but rather presupposed as prior to it and already accepted. The primitive notion of heat was taken as empirically established, especially through calorimetry regarded as a subject in its own right, prior to thermodynamics. Jointly primitive with this notion of heat were the notions of empirical temperature and thermal equilibrium. This framework also took as primitive the notion of transfer of energy as work. This framework did not presume a concept of energy in general, but regarded it as derived or synthesized from the prior notions of heat and work. By one author, this framework has been called the "thermodynamic" approach.

The first explicit statement of the first law of thermodynamics, by Rudolf Clausius in 1850, referred to cyclic thermodynamic processes.

> *In all cases in which work is produced by the agency of heat, a quantity of heat is consumed which is proportional to the work done; and conversely, by the expenditure of an equal quantity of work an equal quantity of heat is produced.*

Clausius also stated the law in another form, referring to the existence of a function of state of the system, the internal energy, and expressed it in terms of a differential equation for the increments of a thermodynamic process. This equation may be described as follows:

> *In a thermodynamic process involving a closed system, the increment in the internal energy is equal to the difference between the heat accumulated by the system and the work done by it.*

Because of its definition in terms of increments, the value of the internal energy of a system is not uniquely defined. It is defined only up to an arbitrary additive constant of integration, which can be adjusted to give arbitrary reference zero levels. This non-uniqueness is in keeping with the abstract mathematical nature of the internal energy. The internal energy is customarily stated relative to a conventionally chosen standard reference state of the system.

The concept of internal energy is considered by Bailyn to be of "enormous interest". Its quantity cannot be immediately measured, but can only be inferred, by differencing actual immediate measurements. Bailyn likens it to the energy states of an atom, that were revealed by Bohr's energy relation $h\nu = E_{n''} - E_{n'}$. In each case, an unmeasurable quantity (the internal energy, the atomic energy level) is revealed by considering the difference of measured quantities (increments of internal energy, quantities of emitted or absorbed radiative energy).

Conceptual Revision: the "Mechanical Approach"

In 1907, George H. Bryan wrote about systems between which there is no transfer of matter (closed systems): "Definition. When energy flows from one system or part of a system to another otherwise than by the performance of mechanical work, the energy so transferred is called *heat*." This definition may be regarded as expressing a conceptual revision, as follows. This was systematically expounded in 1909 by Constantin Carathéodory, whose attention had been drawn to it by Max Born. Largely through Born's influence, this revised conceptual approach to the definition of heat came to be preferred by many twentieth-century writers. It might be called the "mechanical approach".

Energy can also be transferred from one thermodynamic system to another in association with transfer of matter. Born points out that in general such energy transfer is not resolvable uniquely into work and heat moieties. In general, when there is transfer of energy associated with matter transfer, work and heat transfers can be distinguished only when they pass through walls physically separate from those for matter transfer.

The "mechanical" approach postulates the law of conservation of energy. It also postulates that energy can be transferred from one thermodynamic system to another adiabatically as work, and that energy can be held as the internal energy of a thermodynamic system. It also postulates that energy can be transferred from one thermodynamic system to another by a path that is non-adiabatic, and is unaccompanied by matter transfer. Initially, it "cleverly" (according to Bailyn) refrains from labelling as 'heat' such non-adiabatic, unaccompanied transfer of energy. It rests on the primitive notion of *walls*, especially adiabatic walls and non-adiabatic walls, defined as follows. Temporarily, only for purpose of this definition, one can prohibit transfer of energy as work across a wall of interest. Then walls of interest fall into two classes, (a) those such that arbitrary systems separated by them remain independently in their own previously established respective states of internal thermodynamic equilibrium; they are defined as adiabatic; and (b) those without such independence; they are defined as non-adiabatic.

This approach derives the notions of transfer of energy as heat, and of temperature, as theoretical developments, not taking them as primitives. It regards calorimetry as a derived theory. It has an early origin in the nineteenth century, for example in the work of Helmholtz, but also in the work of many others.

Conceptually Revised Statement, According to the Mechanical Approach

The revised statement of the first law postulates that a change in the internal energy of a system due to any arbitrary process, that takes the system from a given initial thermodynamic state to a given final equilibrium thermodynamic state, can be determined through the physical existence, for those given states, of a reference process that occurs purely through stages of adiabatic work.

The revised statement is then

> *For a closed system, in any arbitrary process of interest that takes it from an initial to a final state of internal thermodynamic equilibrium, the change of internal energy is the same as that for a reference adiabatic work process that links those two states. This is so regardless of the path of the process of interest, and regardless of whether it is an adiabatic or a non-adiabatic process. The reference adiabatic work process may be chosen arbitrarily from amongst the class of all such processes.*

This statement is much less close to the empirical basis than are the original statements, but is often regarded as conceptually parsimonious in that it rests only on the concepts of adiabatic work and of non-adiabatic processes, not on the concepts of transfer of energy as heat and of empirical temperature that are presupposed by the original statements. Largely through the influence of Max Born, it is often regarded as theoretically preferable because of this conceptual parsimony. Born particularly observes that the revised approach avoids thinking in terms of what he calls the "imported engineering" concept of heat engines.

Basing his thinking on the mechanical approach, Born in 1921, and again in 1949, proposed to revise the definition of heat. In particular, he referred to the work of Constantin Carathéodory, who had in 1909 stated the first law without defining quantity of heat. Born's definition was specifically for transfers of energy without transfer of matter, and it has been widely followed in textbooks (examples:). Born observes that a transfer of matter between two systems is accompanied by a transfer of internal energy that cannot be resolved into heat and work components. There can be pathways to other systems, spatially separate from that of the matter transfer, that allow heat and work transfer independent of and simultaneous with the matter transfer. Energy is conserved in such transfers.

Description

The first law of thermodynamics for a closed system was expressed in two ways by Clausius. One way referred to cyclic processes and the inputs and outputs of the system, but did not refer to increments in the internal state of the system. The other way referred to an incremental change in the internal state of the system, and did not expect the process to be cyclic.

A cyclic process is one that can be repeated indefinitely often, returning the system to its initial state. Of particular interest for single cycle of a cyclic process are the net work done, and the net heat taken in (or 'consumed', in Clausius' statement), by the system.

In a cyclic process in which the system does net work on its surroundings, it is observed to be physically necessary not only that heat be taken into the system, but also, importantly, that some heat leave the system. The difference is the heat converted by the cycle into work. In each repetition of a cyclic process, the net work done by the system, measured in mechanical units, is proportional to the heat consumed, measured in calorimetric units.

The constant of proportionality is universal and independent of the system and in 1845 and 1847 was measured by James Joule, who described it as the *mechanical equivalent of heat.*

In a non-cyclic process, the change in the internal energy of a system is equal to net energy added as heat to the system minus the net work done by the system, both being measured in mechanical units. Taking ΔU as a change in internal energy, one writes

$$\Delta U = Q - W$$

where Q denotes the net quantity of heat supplied to the system by its surroundings and W denotes the net work done by the system. This sign convention is implicit in Clausius' statement of the law given above. It originated with the study of heat engines that produce useful work by consumption of heat.

Often nowadays, however, writers use the IUPAC convention by which the first law is formulated with work done on the system by its surroundings having a positive sign. With this now often used sign convention for work, the first law for a closed system may be written:

$$\Delta U = Q + W \text{ (sign convention of IUPAC).}$$

This convention follows physicists such as Max Planck, and considers all net energy transfers to the system as positive and all net energy transfers from the system as negative, irrespective of any use for the system as an engine or other device.

When a system expands in a fictive quasistatic process, the work done by the system on the environment is the product, $P\,dV$, of pressure, P, and volume change, dV, whereas the work done *on* the system is $-P\,dV$. Using either sign convention for work, the change in internal energy of the system is:

$$dU = \delta Q - P dV \text{ (quasi-static process),}$$

where δQ denotes the infinitesimal increment of heat supplied to the system from its surroundings.

Work and heat are expressions of actual physical processes of supply or removal of energy, while the internal energy U is a mathematical abstraction that keeps account of the exchanges of energy that befall the system. Thus the term heat for Q means "that amount of energy added or removed by conduction of heat or by thermal radiation", rather than referring to a form of energy within the system. Likewise, the term work energy for W means "that amount of energy gained or lost as the result of work". Internal energy is a property of the system whereas work done and heat supplied are not. A significant result of this distinction is that a given internal energy change ΔU can be achieved by, in principle, many combinations of heat and work.

Various Statements of the Law for Closed Systems

The law is of great importance and generality and is consequently thought of from several points of view. Most careful textbook statements of the law express it for closed systems. It is stated in several ways, sometimes even by the same author.

For the thermodynamics of closed systems, the distinction between transfers of energy as work and as heat is central and is within the scope of the present article. For the thermodynamics of open systems, such a distinction is beyond the scope of the present article, but some limited comments are made on it in the section below headed 'First law of thermodynamics for open systems'.

There are two main ways of stating a law of thermodynamics, physically or mathematically. They should be logically coherent and consistent with one another.

An example of a physical statement is that of Planck (1897/1903):

> It is in no way possible, either by mechanical, thermal, chemical, or other devices, to obtain perpetual motion, i.e. it is impossible to construct an engine which will work in a cycle and produce continuous work, or kinetic energy, from nothing.

This physical statement is restricted neither to closed systems nor to systems with states that are strictly defined only for thermodynamic equilibrium; it has meaning also for open systems and for systems with states that are not in thermodynamic equilibrium.

An example of a mathematical statement is that of Crawford (1963):

> For a given system we let ΔE^{kin} = large-scale mechanical energy, ΔE^{pot} = large-scale potential energy, and ΔE^{tot} = total energy. The first two quantities are specifiable in terms of appropriate mechanical variables, and by definition
>
> $$E^{\mathrm{tot}} = E^{\mathrm{kin}} + E^{\mathrm{pot}} + U.$$
>
> For any finite process, whether reversible or irreversible,
>
> $$\Delta E^{\mathrm{tot}} = \Delta E^{\mathrm{kin}} + \Delta E^{\mathrm{pot}} + \Delta U.$$

> The first law in a form that involves the principle of conservation of energy more generally is
>
> $$\Delta E^{tot} = Q + W.$$
>
> Here Q and W are heat and work added, with no restrictions as to whether the process is reversible, quasistatic, or irreversible.[Warner, *Am. J. Phys.*, **29**, 124 (1961)]

This statement by Crawford, for W, uses the sign convention of IUPAC, not that of Clausius. Though it does not explicitly say so, this statement refers to closed systems, and to internal energy U defined for bodies in states of thermodynamic equilibrium, which possess well-defined temperatures.

The history of statements of the law for closed systems has two main periods, before and after the work of Bryan (1907), of Carathéodory (1909), and the approval of Carathéodory's work given by Born (1921). The earlier traditional versions of the law for closed systems are nowadays often considered to be out of date.

Carathéodory's celebrated presentation of equilibrium thermodynamics refers to closed systems, which are allowed to contain several phases connected by internal walls of various kinds of impermeability and permeability (explicitly including walls that are permeable only to heat). Carathéodory's 1909 version of the first law of thermodynamics was stated in an axiom which refrained from defining or mentioning temperature or quantity of heat transferred. That axiom stated that the internal energy of a phase in equilibrium is a function of state, that the sum of the internal energies of the phases is the total internal energy of the system, and that the value of the total internal energy of the system is changed by the amount of work done adiabatically on it, considering work as a form of energy. That article considered this statement to be an expression of the law of conservation of energy for such systems. This version is nowadays widely accepted as authoritative, but is stated in slightly varied ways by different authors.

Such statements of the first law for closed systems assert the existence of internal energy as a function of state defined in terms of adiabatic work. Thus heat is not defined calorimetrically or as due to temperature difference. It is defined as a residual difference between change of internal energy and work done on the system, when that work does not account for the whole of the change of internal energy and the system is not adiabatically isolated.

The 1909 Carathéodory statement of the law in axiomatic form does not mention heat or temperature, but the equilibrium states to which it refers are explicitly defined by variable sets that necessarily include "non-deformation variables", such as pressures, which, within reasonable restrictions, can be rightly interpreted as empirical temperatures, and the walls connecting the phases of the system are explicitly defined as possibly impermeable to heat or permeable only to heat.

According to Münster (1970), "A somewhat unsatisfactory aspect of Carathéodory's theory is that a consequence of the Second Law must be considered at this point [in the statement of the first law], i.e. that it is not always possible to reach any state 2 from any other state 1 by means of an adiabatic process." Münster instances that no adiabatic process can reduce the internal energy of a system at constant volume. Carathéodory's paper asserts that its statement of the first law corresponds exactly to Joule's experimental arrangement, regarded as an instance of adiabatic work. It does not point out that Joule's experimental arrangement performed essentially irreversible work, through friction of paddles in a liquid, or passage of electric current through a resistance inside the system, driven by motion of a coil and inductive heating, or by an external current source, which can access the system only by the passage of electrons, and so is not strictly adiabatic, because electrons are a form of matter, which cannot penetrate adiabatic walls. The paper goes on to base its main argument on the possibility of quasi-static adiabatic work, which is essentially reversible. The paper asserts that it will avoid reference to Carnot cycles, and then proceeds to base its argument on cycles of forward and backward quasi-static adiabatic stages, with isothermal stages of zero magnitude.

Sometimes the concept of internal energy is not made explicit in the statement.

Sometimes the existence of the internal energy is made explicit but work is not explicitly mentioned in the statement of the first postulate of thermodynamics. Heat supplied is then defined as the residual change in internal energy after work has been taken into account, in a non-adiabatic process.

A respected modern author states the first law of thermodynamics as "Heat is a form of energy", which explicitly mentions neither internal energy nor adiabatic work. Heat is defined as energy transferred by thermal contact with a reservoir, which has a temperature, and is generally so large that addition and removal of heat do not alter its temperature. A current student text on chemistry defines heat thus: "*heat* is the exchange of thermal energy between a system and its surroundings caused by a temperature difference." The author then explains how heat is defined or measured by calorimetry, in terms of heat capacity, specific heat capacity, molar heat capacity, and temperature.

A respected text disregards the Carathéodory's exclusion of mention of heat from the statement of the first law for closed systems, and admits heat calorimetrically defined along with work and internal energy. Another respected text defines heat exchange as determined by temperature difference, but also mentions that the Born (1921) version is "completely rigorous". These versions follow the traditional approach that is now considered out of date, exemplified by that of Planck (1897/1903).

Evidence for the First Law of Thermodynamics for Closed Systems

The first law of thermodynamics for closed systems was originally induced from empirically observed evidence, including calorimetric evidence. It is nowadays, however, taken to provide the definition of heat via the law of conservation of energy and the definition of work in terms of changes in the external parameters of a system. The original discovery of the law was gradual over a period of perhaps half a century or more, and some early studies were in terms of cyclic processes.

The following is an account in terms of changes of state of a closed system through compound processes that are not necessarily cyclic. This account first considers processes for which the first law is easily verified because of their simplicity, namely adiabatic processes (in which there is no transfer as heat) and adynamic processes (in which there is no transfer as work).

Adiabatic Processes

In an adiabatic process, there is transfer of energy as work but not as heat. For all adiabatic process that takes a system from a given initial state to a given final state, irrespective of how the work is done, the respective eventual total quantities of energy transferred as work are one and the same, determined just by the given initial and final states. The work done on the system is defined and measured by changes in mechanical or quasi-mechanical variables external to the system. Physically, adiabatic transfer of energy as work requires the existence of adiabatic enclosures.

For instance, in Joule's experiment, the initial system is a tank of water with a paddle wheel inside. If we isolate the tank thermally, and move the paddle wheel with a pulley and a weight, we can relate the increase in temperature with the distance descended by the mass. Next, the system is returned to its initial state, isolated again, and the same amount of work is done on the tank using different devices (an electric motor, a chemical battery, a spring,...). In every case, the amount of work can be measured independently. The return to the initial state is not conducted by doing adiabatic work on the system. The evidence shows that the final state of the water (in particular, its temperature and volume) is the same in every case. It is irrelevant if the work is electrical, mechanical, chemical,... or if done suddenly or slowly, as long as it is performed in an adiabatic way, that is to say, without heat transfer into or out of the system.

Evidence of this kind shows that to increase the temperature of the water in the tank, the qualitative kind of adiabatically performed work does not matter. No qualitative kind of adiabatic work has ever been observed to decrease the temperature of the water in the tank.

A change from one state to another, for example an increase of both temperature and volume, may be conducted in several stages, for example by externally supplied electrical work on a resistor in the body, and adiabatic expansion allowing the body to do work on the surroundings. It needs to be shown that the time order of the stages, and their relative magnitudes, does not affect the amount of adiabatic work that needs to be done for the change of state. According to one respected scholar: "Unfortunately, it does not seem that experiments of this kind have ever been carried out carefully. ... We must therefore admit that the statement which we have enunciated here, and which is equivalent to the first law of thermodynamics, is not well founded on direct experimental evidence." Another expression of this view is "... no systematic precise experiments to verify this generalization directly have ever been attempted."

This kind of evidence, of independence of sequence of stages, combined with the above-mentioned evidence, of independence of qualitative kind of work, would show the existence of an important state variable that corresponds with adiabatic work, but not that such a state variable represented a conserved quantity. For the latter, another step of evidence is needed, which may be related to the concept of reversibility, as mentioned below.

That important state variable was first recognized and denoted U by Clausius in 1850, but he did not then name it, and he defined it in terms not only of work but also of heat transfer in the same process. It was also independently recognized in 1850 by Rankine, who also denoted it U; and in 1851 by Kelvin who then called it "mechanical energy", and later "intrinsic energy". In 1865, after some hestitation, Clausius began calling his state function U "energy". In 1882 it was named as the *internal energy* by Helmholtz. If only adiabatic processes were of interest, and heat could be ignored, the concept of internal energy would hardly arise or be needed. The relevant physics would be largely covered by the concept of potential energy, as was intended in the 1847 paper of Helmholtz on the principle of conservation of energy, though that did not deal with forces that cannot be described by a potential, and thus did not fully justify the principle. Moreover, that paper was critical of the early work of Joule that had by then been performed. A great merit of the internal energy concept is that it frees thermodynamics from a restriction to cyclic processes, and allows a treatment in terms of thermodynamic states.

In an adiabatic process, adiabatic work takes the system either from a reference state O with internal energy $U(O)$ to an arbitrary one A with internal energy $U(A)$, or from the state A to the state O:

$$U(A) = U(O) - W_{O \to A}^{adiabatic} \text{ or } U(O) = U(A) - W_{A \to O}^{adiabatic}$$

Except under the special, and strictly speaking, fictional, condition of reversibility, only one of the processes $adiabatic, O \to A$ or $adiabatic, A \to O$ is empirically feasible by a simple application of externally supplied work. The reason for this is given as the second law of thermodynamics and is not considered in the present article.

The fact of such irreversibility may be dealt with in two main ways, according to different points of view:

- Since the work of Bryan (1907), the most accepted way to deal with it nowadays, followed by Carathéodory, is to rely on the previously established concept of quasi-static processes, as follows. Actual physical processes of transfer of energy as work are always at least to some degree irreversible. The irreversibility is often due to mechanisms known as dissipative, that transform bulk kinetic energy into internal energy. Examples are friction and viscosity. If the process is performed more slowly, the frictional or viscous dissipation is less. In the limit of infinitely slow performance, the dissipation tends to zero and then the limiting process, though fictional rather than actual, is notionally reversible, and is called quasi-static. Throughout the course of the fictional limiting quasi-static process, the internal intensive variables of the system are equal to the external intensive variables, those that describe the reactive forces exerted by the surroundings. This can be taken to justify the formula

$$W_{A\to O}^{\text{adiabatic, quasi-static}} = -W_{O\to A}^{\text{adiabatic, quasi-static}}.$$

- Another way to deal with it is to allow that experiments with processes of heat transfer to or from the system may be used to justify the formula above. Moreover, it deals to some extent with the problem of lack of direct experimental evidence that the time order of stages of a process does not matter in the determination of internal energy. This way does not provide theoretical purity in terms of adiabatic work processes, but is empirically feasible, and is in accord with experiments actually done, such as the Joule experiments mentioned just above, and with older traditions.

The formula (1) above allows that to go by processes of quasi-static adiabatic work from the state A to the state B we can take a path that goes through the reference state O, since the quasi-static adiabatic work is independent of the path

$$-W_{A\to B}^{\text{adiabatic,quasi-static}} = -W_{A\to O}^{\text{adiabatic,quasi-static}} - W_{O\to B}^{\text{adiabatic,quasi-static}} = W_{O\to A}^{\text{adiabatic,quasi-static}} - W_{O\to B}^{\text{adiabatic,quasi-static}} = -U(A) + U(B) = \Delta U$$

This kind of empirical evidence, coupled with theory of this kind, largely justifies the following statement:

> *For all adiabatic processes between two specified states of a closed system of any nature, the net work done is the same regardless the details of the process, and determines a state function called internal energy,* U *.”*

Adynamic Processes

A complementary observable aspect of the first law is about heat transfer. Adynamic transfer of energy as heat can be measured empirically by changes in the surroundings of the system of interest by calorimetry. This again requires the existence of adiabatic enclosure of the entire process, system and surroundings, though the separating wall between the surroundings and the system is thermally conductive or radiatively permeable, not adiabatic. A calorimeter can rely on measurement of sensible heat, which requires the existence of thermometers and measurement of temperature change in bodies of known sensible heat capacity under specified conditions; or it can rely on the measurement of latent heat, through measurement of masses of material that change phase, at temperatures fixed by the occurrence of phase changes under specified conditions in bodies of known latent heat of phase change. The calorimeter can be calibrated by adiabatically doing externally determined work on it. The most accurate method is by passing an electric current from outside through a resistance inside the calorimeter. The calibration allows comparison of calorimetric measurement of quantity of heat transferred with quantity of energy transferred as work. According to one textbook, "The most common device for measuring ΔU is an adiabatic bomb calorimeter." According to another textbook, "Calorimetry is widely used in present day laboratories." According to one opinion, "Most thermodynamic data come from calorimetry..." According to another opinion, "The most common method of measuring "heat" is with a calorimeter."

When the system evolves with transfer of energy as heat, without energy being transferred as work, in an adynamic process, the heat transferred to the system is equal to the increase in its internal energy:

$$Q_{A\to B}^{\text{adynamic}} = \Delta U.$$

General Case for Reversible Processes

Heat transfer is practically reversible when it is driven by practically negligibly small temperature gradients. Work transfer is practically reversible when it occurs so slowly that there are no frictional effects within the system; frictional effects outside the system should also be zero if the process is to be globally reversible. For a particular reversible process in general, the work done reversibly on the system, $W_{A\to B}^{\text{path}\,P_0,\,\text{reversible}}$, , and the heat transferred reversibly to the system, $Q_{A\to B}^{\text{path}\,P_0,\,\text{reversible}}$ are not required to occur respectively adiabatically or adynamically, but they must belong to the same particular process defined by its particular reversible path, P_0, through the space of thermodynamic states. Then the work and heat transfers can occur and be calculated simultaneously.

Putting the two complementary aspects together, the first law for a particular reversible process can be written

$$-W_{A\to B}^{\text{path}\,P_0,\,\text{reversible}} + Q_{A\to B}^{\text{path}\,P_0,\,\text{reversible}} = \Delta U.$$

This combined statement is the expression the first law of thermodynamics for reversible processes for closed systems.

In particular, if no work is done on a thermally isolated closed system we have

$$\Delta U = 0.$$

This is one aspect of the law of conservation of energy and can be stated:

The internal energy of an isolated system remains constant.

General Case for Irreversible Processes

If, in a process of change of state of a closed system, the energy transfer is not under a practically zero temperature gradient and practically frictionless, then the process is irreversible. Then the heat and work transfers may be difficult to calculate, and irreversible thermodynamics is called for. Nevertheless, the first law still holds and provides a check on the measurements and calculations of the work done irreversibly on the system, $W_{A \to B}^{\mathrm{path}\, P_1,\, \mathrm{irreversible}}$, and the heat transferred irreversibly to the system, $Q_{A \to B}^{\mathrm{path}\, P_1,\, \mathrm{irreversible}}$, which belong to the same particular process defined by its particular irreversible path, P_1, through the space of thermodynamic states.

$$-W_{A \to B}^{\mathrm{path}\, P_1,\, \mathrm{irreversible}} + Q_{A \to B}^{\mathrm{path}\, P_1,\, \mathrm{irreversible}} = \Delta U.$$

This means that the internal energy U is a function of state and that the internal energy change ΔU between two states is a function only of the two states.

Overview of the Weight of Evidence for the Law

The first law of thermodynamics is so general that its predictions cannot all be directly tested. In many properly conducted experiments it has been precisely supported, and never violated. Indeed, within its scope of applicability, the law is so reliably established, that, nowadays, rather than experiment being considered as testing the accuracy of the law, it is more practical and realistic to think of the law as testing the accuracy of experiment. An experimental result that seems to violate the law may be assumed to be inaccurate or wrongly conceived, for example due to failure to account for an important physical factor. Thus, some may regard it as a principle more abstract than a law.

State Functional Formulation for Infinitesimal Processes

When the heat and work transfers in the equations above are infinitesimal in magnitude, they are often denoted by δ, rather than exact differentials denoted by d, as a reminder that heat and work do not describe the *state* of any system. The integral of an inexact differential depends upon the particular path taken through the space of thermodynamic parameters while the integral of an exact differential depends only upon the initial and final states. If the initial and final states are the same, then the integral of an inexact differential may or may not be zero, but the integral of an exact differential is always zero. The path taken by a thermodynamic system through a chemical or physical change is known as a thermodynamic process.

The first law for a closed homogeneous system may be stated in terms that include concepts that are established in the second law. The internal energy U may then be expressed as a function of the system's defining state variables S, entropy, and V, volume: $U = U(S, V)$. In these terms, T, the system's temperature, and P, its pressure, are partial derivatives of U with respect to S and V. These variables are important throughout thermodynamics, though not necessary for the statement of the first law. Rigorously, they are defined only when the system is in its own state of internal thermodynamic equilibrium. For some purposes, the concepts provide good approximations for scenarios sufficiently near to the system's internal thermodynamic equilibrium.

The first law requires that:

$$dU = \delta Q - \delta W \qquad \text{(closed system, general process, quasi-static or irreversible).}$$

Then, for the fictive case of a reversible process, dU can be written in terms of exact differentials. One may imagine reversible changes, such that there is at each instant negligible departure from thermodynamic equilibrium within the system. This excludes isochoric work. Then, mechanical work is given by $\delta W = -P\,dV$ and the quantity of heat added can be expressed as $\delta Q = T\,dS$. For these conditions

$$dU = TdS - PdV \qquad \text{(closed system, reversible process).}$$

While this has been shown here for reversible changes, it is valid in general, as U can be considered as a thermodynamic state function of the defining state variables S and V:

$$dU = TdS - PdV \text{ (closed system, general process, quasi-static or irreversible).}$$

Above equation is known as the fundamental thermodynamic relation for a closed system in the energy representation, for which the defining state variables are S and V, with respect to which T and P are partial derivatives of U. It is only in the fictive reversible case, when isochoric work is excluded, that the work done and heat transferred are given by $-P\,dV$ and TdS.

In the case of a closed system in which the particles of the system are of different types and, because chemical reactions may occur, their respective numbers are not necessarily constant, the fundamental thermodynamic relation for dU becomes:

$$dU = TdS - PdV + \sum_i \mu_i dN_i.$$

where dN_i is the (small) increase in amount of type-i particles in the reaction, and μ_i is known as the chemical potential of the type-i particles in the system. If dN_i is expressed in mol then μ_i is expressed in J/mol. If the system has more external mechanical variables than just the volume that can change, the fundamental thermodynamic relation further generalizes to:

$$dU = TdS - \sum_i X_i dx_i + \sum_j \mu_j dN_j.$$

Here the X_i are the generalized forces corresponding to the external variables x_i. The parameters X_i are independent of the size of the system and are called intensive parameters and the x_i are proportional to the size and called extensive parameters.

For an open system, there can be transfers of particles as well as energy into or out of the system during a process. For this case, the first law of thermodynamics still holds, in the form that the internal energy is a function of state and the change of internal energy in a process is a function only of its initial and final states, as noted in the section below headed First law of thermodynamics for open systems.

A useful idea from mechanics is that the energy gained by a particle is equal to the force applied to the particle multiplied by the displacement of the particle while that force is applied. Now consider the first law without the heating term: $dU = -PdV$. The pressure P can be viewed as a force (and in fact has units of force per unit area) while $dV$is the displacement (with units of distance times area). We may say, with respect to this work term, that a pressure difference forces a transfer of volume, and that the product of the two (work) is the amount of energy transferred out of the system as a result of the process. If one were to make this term negative then this would be the work done on the system.

It is useful to view the TdS term in the same light: here the temperature is known as a "generalized" force (rather than an actual mechanical force) and the entropy is a generalized displacement.

Similarly, a difference in chemical potential between groups of particles in the system drives a chemical reaction that changes the numbers of particles, and the corresponding product is the amount of chemical potential energy transformed in process. For example, consider a system consisting of two phases: liquid water and water vapor. There is a generalized "force" of evaporation that drives water molecules out of the liquid. There is a generalized "force" of condensation that drives vapor molecules out of the vapor. Only when these two "forces" (or chemical potentials) are equal is there equilibrium, and the net rate of transfer zero.

The two thermodynamic parameters that form a generalized force-displacement pair are called "conjugate variables". The two most familiar pairs are, of course, pressure-volume, and temperature-entropy.

Spatially Inhomogeneous Systems

Classical thermodynamics is initially focused on closed homogeneous systems (e.g. Planck 1897/1903), which might be regarded as 'zero-dimensional' in the sense that they have no spatial variation. But it is desired to study also systems with distinct internal motion and spatial inhomogeneity. For such systems, the principle of conservation of energy is expressed in terms not only of internal energy as defined for homogeneous systems, but also in terms of kinetic energy and potential energies of parts of the inhomogeneous system with respect to each other and with respect to long-range external forces. How the total energy of a system is allocated between these three more specific kinds of energy varies according to the purposes of different writers; this is because these components of energy are to some extent mathematical artefacts rather than actually measured physical quantities. For any closed homogeneous component of an inhomogeneous closed system, if E denotes the total energy of that component system, one may write

$$E = E^{kin} + E^{pot} + U$$

where E^{kin} and E^{pot} denote respectively the total kinetic energy and the total potential energy of the component closed homogeneous system, and U denotes its internal energy.

Potential energy can be exchanged with the surroundings of the system when the surroundings impose a force field, such as gravitational or electromagnetic, on the system.

A compound system consisting of two interacting closed homogeneous component subsystems has a potential energy of interaction E_{12}^{pot} between the subsystems. Thus, in an obvious notation, one may write

$$E = E_1^{kin} + E_1^{pot} + U_1 + E_2^{kin} + E_2^{pot} + U_2 + E_{12}^{pot}$$

The quantity E_{12}^{pot} in general lacks an assignment to either subsystem in a way that is not arbitrary, and this stands in the way of a general non-arbitrary definition of transfer of energy as work. On occasions, authors make their various respective arbitrary assignments.

The distinction between internal and kinetic energy is hard to make in the presence of turbulent motion within the system, as friction gradually dissipates macroscopic kinetic energy of localised bulk flow into molecular random motion of molecules that is classified as internal energy. The rate of dissipation by friction of kinetic energy of localised bulk flow into internal energy, whether in turbulent or in streamlined flow, is an important quantity in non-equilibrium thermodynamics. This is a serious difficulty for attempts to define entropy for time-varying spatially inhomogeneous systems.

First Law of Thermodynamics for Open Systems

For the first law of thermodynamics, there is no trivial passage of physical conception from the closed system view to an open system view. For closed systems, the concepts of an adiabatic enclosure and of an adiabatic wall are fundamental. Matter and internal energy cannot permeate or penetrate such a wall. For an open system, there is a wall that allows penetration by matter. In general, matter in diffusive motion carries with it some internal energy, and some microscopic potential energy changes accompany the motion. An open system is not adiabatically enclosed.

There are some cases in which a process for an open system can, for particular purposes, be considered as if it were for a closed system. In an open system, by definition hypothetically or potentially, matter can pass between the system and its surroundings. But when, in a particular case, the process of interest involves only hypothetical or potential but no actual passage of matter, the process can be considered as if it were for a closed system.

Internal Energy for an Open System

Since the revised and more rigorous definition of the internal energy of a closed system rests upon the possibility of processes by which adiabatic work takes the system from one state to another, this leaves a problem for the definition of internal energy for an open system, for which adiabatic work is not in general possible. According to Max Born, the transfer of matter and energy across an open connection "cannot be reduced to mechanics". In contrast to the case of closed systems, for open systems, in the presence of diffusion, there is no unconstrained and unconditional physical distinction between convective transfer of internal energy by bulk flow of matter, the transfer of internal energy without transfer of matter (usually called heat conduction and work transfer), and change of various potential energies. The older traditional way and the conceptually revised (Carathéodory) way agree that there is no physically unique definition of heat and work transfer processes between open systems.

In particular, between two otherwise isolated open systems an adiabatic wall is by definition impossible. This problem is solved by recourse to the principle of conservation of energy. This principle allows a composite isolated system to be derived from two other component non-interacting isolated systems, in such a way that the total energy of the composite isolated system is equal to the sum of the total energies of the two component isolated systems. Two previously isolated systems can be subjected to the thermodynamic operation of placement between them of a wall permeable to matter and energy, followed by a time for establishment of a new thermodynamic state of internal equilibrium in the new single unpartitioned system. The internal energies of the initial two systems and of the final new system, considered respectively as closed systems as above, can be measured. Then the law of conservation of energy requires that

$$\Delta U_s + \Delta U_o = 0,$$

where ΔU_s and ΔU_0 denote the changes in internal energy of the system and of its surroundings respectively. This is a statement of the first law of thermodynamics for a transfer between two otherwise isolated open systems, that fits well with the conceptually revised and rigorous statement of the law stated above.

For the thermodynamic operation of adding two systems with internal energies U_1 and U_2, to produce a new system with internal energy U, one may write $U = U_1 + U_2$; the reference states for U, U_1 and U_2 should be specified accordingly, maintaining also that the internal energy of a system be proportional to its mass, so that the internal energies are extensive variables.

There is a sense in which this kind of additivity expresses a fundamental postulate that goes beyond the simplest ideas of classical closed system thermodynamics; the extensivity of some variables is not obvious, and needs explicit expression; indeed one author goes so far as to say that it could be recognized as a fourth law of thermodynamics, though this is not repeated by other authors.

Also of course

$$\Delta N_s + \Delta N_o = 0,$$

where ΔN_s and ΔN_o denote the changes in mole number of a component substance of the system and of its surroundings respectively. This is a statement of the law of conservation of mass.

Process of Transfer of Matter between an Open System and its Surroundings

A system connected to its surroundings only through contact by a single permeable wall, but otherwise isolated, is an open system. If it is initially in a state of contact equilibrium with a surrounding subsystem, a thermodynamic process of transfer of matter can be made to occur between them if the surrounding subsystem is subjected to some thermodynamic operation, for example, removal of a partition between it and some further surrounding subsystem. The removal of the partition in the surroundings initiates a process of exchange between the system and its contiguous surrounding subsystem.

An example is evaporation. One may consider an open system consisting of a collection of liquid, enclosed except where it is allowed to evaporate into or to receive condensate from its vapor above it, which may be considered as its contiguous surrounding subsystem, and subject to control of its volume and temperature.

A thermodynamic process might be initiated by a thermodynamic operation in the surroundings, that mechanically increases in the controlled volume of the vapor. Some mechanical work will be done within the surroundings by the vapor, but also some of the parent liquid will evaporate and enter the vapor collection which is the contiguous surrounding subsystem. Some internal energy will accompany the vapor that leaves the system, but it will not make sense to try to uniquely identify part of that internal energy as heat and part of it as work. Consequently, the energy transfer that accompanies the transfer of matter between the system and its surrounding subsystem cannot be uniquely split into heat and work transfers to or from the open system. The component of total energy transfer that accompanies the transfer of vapor into the surrounding subsystem is customarily called 'latent heat of evaporation', but this use of the word heat is a quirk of customary historical language, not in strict compliance with the thermodynamic definition of transfer of energy as heat. In this example, kinetic energy of bulk flow and potential energy with respect to long-range external forces such as gravity are both considered to be zero. The first law of thermodynamics refers to the change of internal energy of the open system, between its initial and final states of internal equilibrium.

Open System with Multiple Contacts

An open system can be in contact equilibrium with several other systems at once.

This includes cases in which there is contact equilibrium between the system, and several subsystems in its surroundings, including separate connections with subsystems through walls that are permeable to the transfer of matter and internal energy as heat and allowing friction of passage of the transferred matter, but immovable, and separate connections through adiabatic walls with others, and separate connections through diathermic walls impermeable to matter with yet others. Because there are physically separate connections that are permeable to energy but impermeable to matter, between the system and its surroundings, energy transfers between them can occur with definite heat and work characters. Conceptually essential here is that the internal energy transferred with the transfer of matter is measured by a variable that is mathematically independent of the variables that measure heat and work.

With such independence of variables, the total increase of internal energy in the process is then determined as the sum of the internal energy transferred from the surroundings with the transfer of matter through the walls that are permeable to it, and of the internal energy transferred to the system as heat through the diathermic walls, and of the energy transferred to the system as work through the adiabatic walls, including the energy transferred to the system by long-range forces. These simultaneously transferred quantities of energy are defined by events in the surroundings of the system. Because the internal energy transferred with matter is not in general uniquely resolvable into heat and work components, the total energy transfer cannot in general be uniquely resolved into heat and work components. Under these conditions, the following formula can describe the process in terms of externally defined thermodynamic variables, as a statement of the first law of thermodynamics:

$$\Delta U_0 = Q - W - \sum_{i=1}^{m} \Delta U_i \quad \text{(suitably defined surrounding subsystems, general process, quasi-static or irreversible),}$$

where ΔU_0 denotes the change of internal energy of the system, and ΔU_i denotes the change of internal energy of the ith of the m surrounding subsystems that are in open contact with the system, due to transfer between the system and that ith surrounding subsystem, and Q denotes the internal energy transferred as heat from the heat reservoir of the surroundings to the system, and W denotes the energy transferred from the system to the surrounding subsystems that are in adiabatic connection with it. The case of a wall that is permeable to matter and can move so as to allow transfer of energy as work is not considered here.

Combination of First and Second Laws

If the system is described by the energetic fundamental equation, $U_0 = U_0(S, V, N_j)$, and if the process can be described in the quasi-static formalism, in terms of the internal state variables of the system, then the process can also be described by a combination of the first and second laws of thermodynamics, by the formula

$$dU_0 = TdS - PdV + \sum_{j=1}^{n} \mu_j dN_j$$

where there are n chemical constituents of the system and permeably connected surrounding subsystems, and where T, S, P, V, N_j, and μ_j, are defined as above.

For a general natural process, there is no immediate term-wise correspondence between above given two equations, because they describe the process in different conceptual frames.

Nevertheless, a conditional correspondence exists. There are three relevant kinds of wall here: purely diathermal, adiabatic, and permeable to matter. If two of those kinds of wall are sealed off, leaving only one that permits transfers of energy, as work, as heat, or with matter, then the remaining permitted terms correspond precisely. If two of the kinds of wall are left unsealed, then energy transfer can be shared between them, so that the two remaining permitted terms do not correspond precisely.

For the special fictive case of quasi-static transfers, there is a simple correspondence. For this, it is supposed that the system has multiple areas of contact with its surroundings. There are pistons that allow adiabatic work, purely diathermal walls, and open connections with surrounding subsystems of completely controllable chemical potential (or equivalent controls for charged species). Then, for a suitable fictive quasi-static transfer, one can write

$$\delta Q = TdS \text{ and } \delta W = PdV \quad \text{(suitably defined surrounding subsystems, quasi-static transfers of energy).}$$

For fictive quasi-static transfers for which the chemical potentials in the connected surrounding subsystems are suitably controlled, these can be put into earlier equation to yield

$$dU_0 = \delta Q - \delta W + \sum_{j=1}^{n} \mu_j dN_j \quad \text{(suitably defined surrounding subsystems, quasi-static transfers).}$$

The reference does not actually write above equation, but what it does write is fully compatible with it. Another helpful account is given by Tschoegl.

There are several other accounts of this, in apparent mutual conflict.

Non-equilibrium Transfers

The transfer of energy between an open system and a single contiguous subsystem of its surroundings is considered also in non-equilibrium thermodynamics. The problem of definition arises also in this case. It may be allowed that the wall between the system and the subsystem is not only permeable to matter and to internal energy, but also may be movable so as to allow work to be done when the two systems have different pressures. In this case, the transfer of energy as heat is not defined.

Methods for study of non-equilibrium processes mostly deal with spatially continuous flow systems. In this case, the open connection between system and surroundings is usually taken to fully surround the system, so that there are no separate connections impermeable to matter but permeable to heat. Except for the special case mentioned above when there is no actual transfer of matter, which can be treated as if for a closed system, in strictly defined thermodynamic terms, it follows that transfer of energy as heat is not defined. In this sense, there is no such thing as 'heat flow' for a continuous-flow open system. Properly, for closed systems, one speaks of transfer of internal energy as heat, but in general, for open systems, one can speak safely only of transfer of internal energy. A factor here is that there are often cross-effects between distinct transfers, for example that transfer of one substance may cause transfer of another even when the latter has zero chemical potential gradient.

Usually transfer between a system and its surroundings applies to transfer of a state variable, and obeys a balance law, that the amount lost by the donor system is equal to the amount gained by the receptor system. Heat is not a state variable. For his 1947 definition of "heat transfer" for discrete open systems, the author Prigogine carefully explains at some length that his definition of it does not obey a balance law. He describes this as paradoxical.

The situation is clarified by Gyarmati, who shows that his definition of "heat transfer", for continuous-flow systems, really refers not specifically to heat, but rather to transfer of internal energy, as follows. He considers a conceptual small cell in a situation of continuous-flow as a system defined in the so-called Lagrangian way, moving with the local center of mass. The flow of matter across the boundary is zero when considered as a flow of total mass. Nevertheless, if the material constitution is of several chemically distinct components that can diffuse with respect to one another, the system is considered to be open, the diffusive flows of the components being defined with respect to the center of mass of the system, and balancing one another as to mass transfer. Still there can be a distinction between bulk flow of internal energy and diffusive flow of internal energy in this case, because the internal energy density does not have to be constant per unit mass of material, and allowing for non-conservation of internal energy because of local conversion of kinetic energy of bulk flow to internal energy by viscosity.

Gyarmati shows that his definition of "the heat flow vector" is strictly speaking a definition of flow of internal energy, not specifically of heat, and so it turns out that his use here of the word heat is contrary to the strict thermodynamic definition of heat, though it is more or less compatible with historical custom, that often enough did not clearly distinguish between heat and internal energy; he writes "that this relation must be considered to be the exact definition of the concept of heat flow, fairly loosely used in experimental physics and heat technics." Apparently in a different frame of thinking from that of the above-mentioned paradoxical usage in the earlier sections of the historic 1947 work by Prigogine, about discrete systems, this usage of Gyarmati is consistent with the later sections of the same 1947 work by Prigogine, about continuous-flow systems, which use the term "heat flux" in just this way. This usage is also followed by Glansdorff and Prigogine in their 1971 text about continuous-flow systems. They write: "Again the flow of internal energy may be split into a convection flow $\rho u\mathbf{v}$ and a conduction flow. This conduction flow is by definition the heat flow $\mathbf{W}$. Therefore: $\mathbf{j}[U] = \rho u\mathbf{v} + \mathbf{W}$ where u denotes the [internal] energy per unit mass. [These authors actually use the symbols E and e to denote internal energy but their notation has been changed here to accord with the notation of the present article. These authors actually use the symbol U to refer to total energy, including kinetic energy of bulk flow.]" This usage is followed also by other writers on non-equilibrium thermodynamics such as Lebon, Jou, and Casas-Vásquez, and de Groot and Mazur. This usage is described by Bailyn as stating the non-convective flow of internal energy, and is listed as his definition number 1, according to the first law of thermodynamics. This usage is also followed by workers in the kinetic theory of gases. This is not the *ad hoc* definition of "reduced heat flux" of Haase.

In the case of a flowing system of only one chemical constituent, in the Lagrangian representation, there is no distinction between bulk flow and diffusion of matter. Moreover, the flow of matter is zero into or out of the cell that moves with the local center of mass. In effect, in this description, one is dealing with a system effectively closed to the transfer of matter. But still one can validly talk of a distinction between bulk flow and diffusive flow of internal energy, the latter driven by a temperature gradient within the flowing material, and being defined with respect to the local center of mass of the bulk flow. In this case of a virtually closed system, because of the zero matter transfer, as noted above, one can safely distinguish between transfer of energy as work, and transfer of internal energy as heat.

A series of Experimants carried out by Joule between 1843 and 1848 from the basis for the First Law of Thermodyanmics

The following are the observations during the Paddle Wheel experiment shown in Figure.

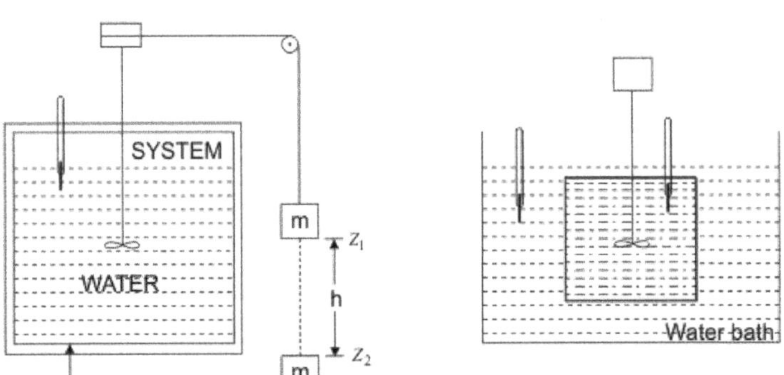

- Work done on the system by lowering the mass m through $(Z_1 - Z_2)$ = change in PE of m

- Temperature of the system was found to increase
- System was brought into contact with a water bath
- System was allowed to come back to initial state
- Energy is transferred as heat from the system to the bath

The system thus executes a cycle which consists of work input to the system followed by the transfer of heat from the system.

$$\underset{\substack{\text{from the} \\ \text{system}}}{\oint dQ} = \underset{\substack{\text{on the} \\ \text{system}}}{\oint dW}$$

Whenever a system undergoes a cyclic change, however complex the cycle may be, the algebraic sum of the work transfer is equal to the algebraic sum of the energy transfer as heat (FIRST LAW OF THERMODYNAMICS).

Sign convention followed in this text:

- Work done by a system on its surroundings is treated as a positive quantity.
- Energy transfer as heat to a system from its surroundings is treated as a positive quantity

$$-\oint dQ - \oint(-dW) = \oint dQ - \oint dW = 0$$

or,

$$\oint (dQ - dW) = 0$$

Heat is Path Function

Lets us consider following two cycles: 1a2b1 and la2cl and apply the first law of thermo-dynamics in above equation to get

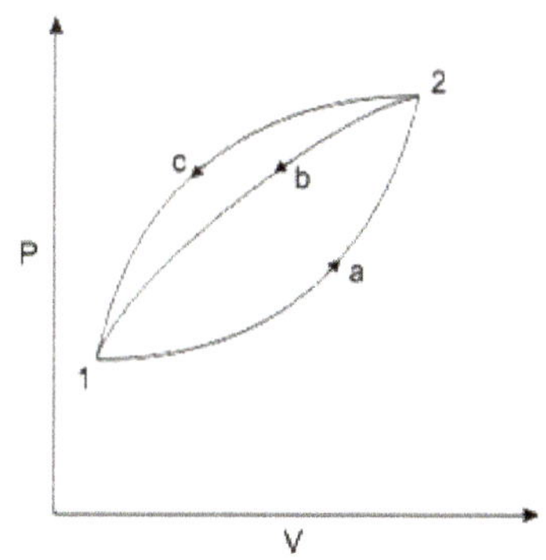

$$\int_{la2} dQ + \int_{2bl} dQ - \int_{la2} dW - \int_{2bl} dW = 0$$

or,

$$\int_{la2} dQ + \int_{2cl} dQ - \int_{la2} dW - \int_{2cl} dW = 0$$

Subtracting Eq. $\int_{la2} dQ + \int_{2cl} dQ - \int_{la2} dW - \int_{2cl} dW = 0$ from Eq. $\int_{la2} dQ + \int_{2bl} dQ - \int_{la2} dW - \int_{2bl} dW = 0$

$$\int_{2bl} dQ - \int_{2cl} dQ - \left(\int_{2bl} dW - \int_{2cl} dW\right) = 0$$

Since, work depends on the path

$$\int_{2bl} dW - \int_{2cl} dW \neq 0$$

Therefore,

$$\int_{2bl} dQ \neq \int_{2cl} dQ$$

Energy transfer as heat is not a point function, neither is it a property of the system.

Heat interaction is a path function.

Energy is a Property of the System

Refer to Figure again and consider Eq. $\int_{2bl} dQ - \int_{2cl} dQ - \left(\int_{2bl} dW - \int_{2cl} dW\right) = 0$

$$\int_{2bl} dQ - \int_{2cl} dQ - \left(\int_{2bl} dW - \int_{2cl} dW\right) = 0$$

$$\int_{2bl} (dQ - dW) = \int_{2cl} (dQ - dW)$$

$\int dQ$ and $\int dW$ depend on path function followed by the system. The quantity $\int (dQ - dW)$ s the same for both the processes $2bl$ and $2cl$ connecting the states 2 and 1. The quantity $\int (dQ - dW)$ does not depend on path followed by the system, depends on the initial and final states. Hence $\int (dQ - dW)$ is an exact differential.

Differential of Property of the System

This property is the internal energy of system, E

$$dU = dQ - dW$$

Energy of an Isolated System

An isolated system is one in which there is no interaction of the system with the surroundings. For an isolated system $dQ = 0, dW = 0$

$$dU = 0 \quad or, \quad U_2 = U_1$$

A Perpetual Motion Machine of First Kind

Thermodynamics originated as a result of man's endeavour to convert the disorganized form of energy (internal energy) into organized form of energy (work).

$$\oint (dQ - dW) = 0 \qquad or, \qquad Q = W$$

An imaginary device which would produce work continuously without absorbing any energy from its surroundings is called a Perpetual Motion Machine of the First kind, (PMMFK). A PMMFK is a device which violates the first law of thermodynamics. It is impossible to devise a PMMFK

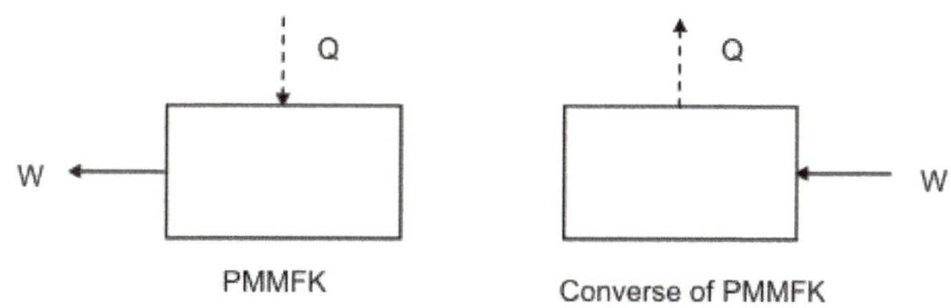

The converse of the above statement is also true, i.e., there can be no machine which would continuously consume work without some other form of energy appearing simultaneously.

Analysis of Closed System

Let us consider a system that refers to a definite quantity of matter which remains constant while the system undergoes a change of state. We shall discuss the following elementary processes involving the closed systems.

Constant Volume Process

Our system is a gas confined in a rigid container of volume V

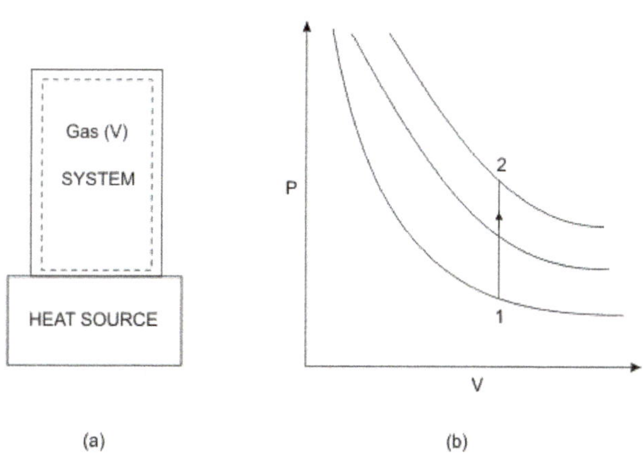

- Let the system be brought into contact with a heat source.
- The energy is exchanged reversibly. The expansion work done (PdV) by the system is zero.
- Applying the first law of thermodynamics, we get

$$dU = dQ$$

or

$$Q = U_2 = U_1$$

- Hence the heat interaction is equal to the change in the internal energy of the system.

Constant Volume Adiabatic Process

Refer to Figure where a change in the state of the system is brought about by performing paddle wheel work on the system.

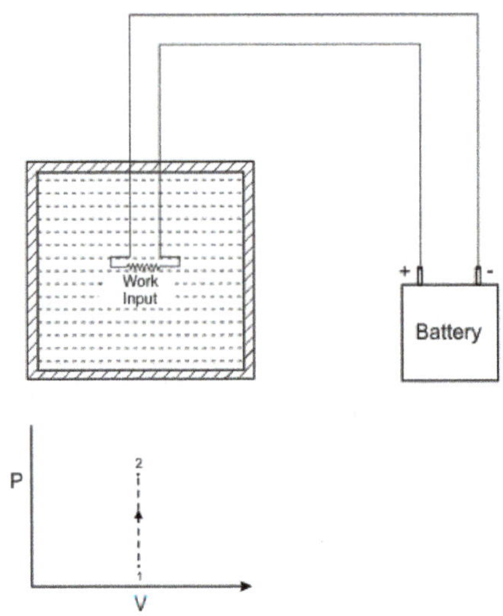

(a)

(b)

The process is irreversible. However, the first law gives

$$dU = -dW$$

or

$$-W = U_2 = U_1$$

Interaction of heat and irreversible work with the system is same in nature. (-W) represents the work done on the system by the surroundings

Specific Heat at Constant Volume

By definition it is the amount of energy required to change the temperature of a unit mass of the substance by one degree.

$$c_v = \left(\frac{dq}{dT}\right)_u$$

While the volume is held constant. For a constant volume process, first law of thermodynamics gives

$$dU = dQ \quad or, \quad du = dq$$

Therefore,

$$c_v = \left(\frac{\partial u}{\partial T}\right)_v$$

Where u is the specific internal energy of the system. If c_v varies with temperature, one can use mean specific heat at constant volume

$$\bar{c}_u = \left[\int_{T_1}^{T_2} c_v dT\right] / (T_2 - T_1)$$

The total quantity of energy transferred during a constant volume process when the system temperature changes from T_1 to T_2

$$Q = m\int_{T_1}^{T_2} c_v dT = U_2 - U_1$$

The unit for c_v is kJ/kgK. The unit of molar specific heat is kJ/kmolK.

Constant Pressure Process

A gas in the piston-cylinder assembly is considered as the system of interest

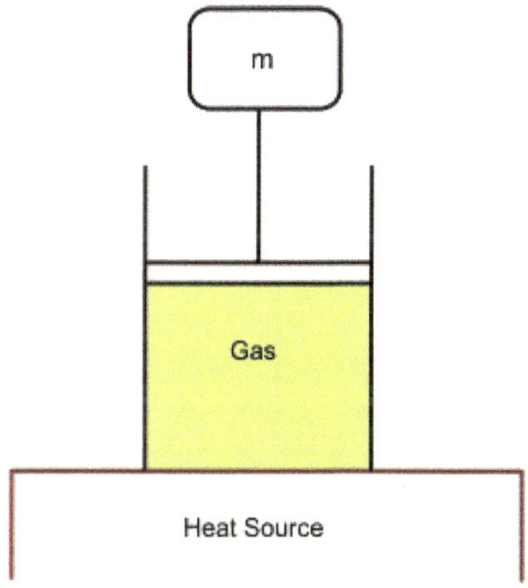

- The pressure is maintained at a constant value by loading the system with a mass.
- The cylinder is brought into contact with a heat source.

Energy transfer as heat takes place reversibly. The work is done by system when it changes from the initial state to the final state.

$$W = \int_1^2 PdV = P(V_2 - V_1)$$

Applying the first law, we get

$$dU = dQ - dW = dQ - PdV = dQ - d(PV)$$

or

$$dQ = dU + d(PV) = d(U + PV)$$

or

$$dQ = dH \quad or, \quad Q = H_2 - H_1$$

The quantity $(U + PV)$ is known as enthalpy, H (a property) of the system. The specific enthalpy h is defined as

$$h = H/m = u + Pv$$

The molar enthalpy is $\hat{h} = H/N$, where N is the mole number of the substance.

Let us assume paddle wheel work is done on the system figure. Also, consider adiabatic walls, so that

$$dQ = 0$$

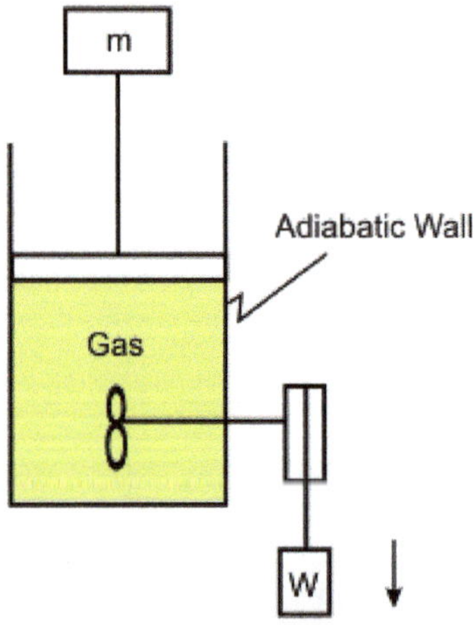

Now the application of first law enables us to write

$$dU = dQ - dW = dQ - (PdV + W_{sh})$$

or

$$-dW_{sh} = dU + PdV = d(U + PV) = dH$$

or

$$-W_{sh} = H_2 - H_1$$

Therefore, the increase in the enthalpy of the system is equal to the amount of shaft work done on the system.

Specific Heat at Constant Pressure

Let us focus on Figure

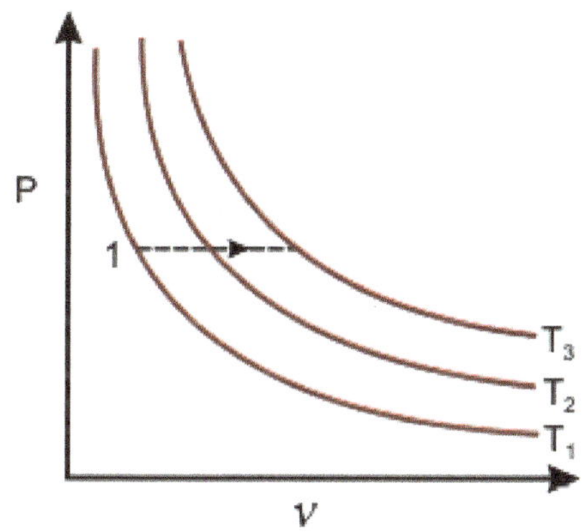

- System changes its state from 1 to 2 following a constant pressure process.
- There will be an accompanying change in temperature.

Specific heat at constant pressure is defined as the quantity of energy required to change the temperature of a unit mass of the substance by one degree during a constant pressure process.

$$c_p = \left(\frac{dq}{dT}\right)_p = \left(\frac{\partial h}{\partial T}\right)_p$$

The total heat interaction for a change in temperature from T_1 to T_2 can be calculated from

$$Q = m\int_{T1}^{T2} c_p dT = H_2 - H_1$$

The molar specific heat at constant pressure can be defined as

$$\hat{c}_p = \left(\frac{\partial \hat{h}}{\partial T}\right)_p$$

Constant Temperature Process

Let us refer to figure

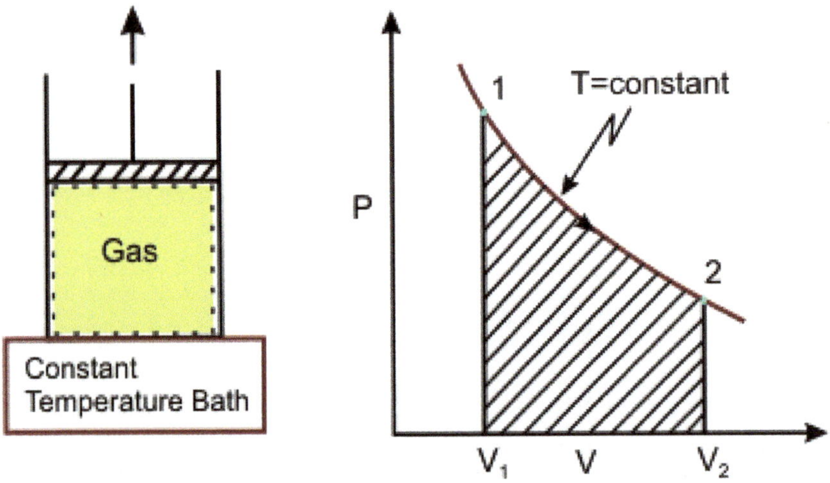

The system is allowed to undergo an expansion process while in contact with constant temperature bath. During the expansion process, the opposing force is continuously reduced. System is in equilibrium at all times. Applying the first law, one can write

$$dU = dQ - dW = dQ - PdV$$

For an ideal gas, the desired property relations are

$$Pv = RT \quad and \quad u = u(T)$$

Since the temperature is held constant, du=0 and Q=W. We can also write

$$q = w = \int_1^2 Pdv = \int_1^2 \frac{RT}{v} dv = RT \ln \frac{v_2}{v_1}$$

Adiabatic Process

1. Irreversible Adiabatic Process

A process in which there is no energy transfer as heat across the boundaries of the system, is called an adiabatic process. For an adiabatic process, Q=0. Paddle wheel work is performed on the system.

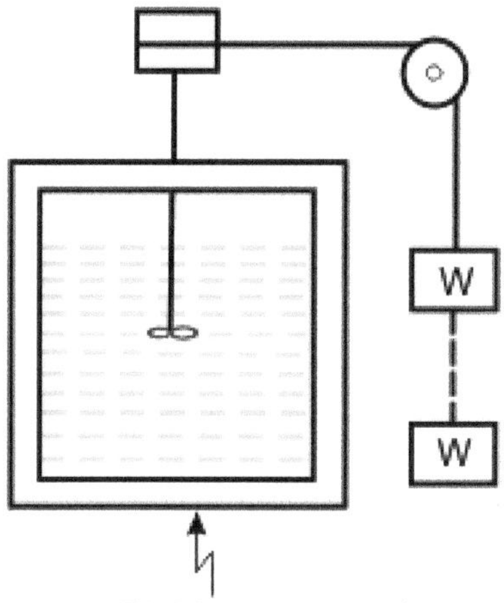

Application of first law gives

$$dU = dQ - dW = dQ - (W) = -dW$$

or

$$\Delta U = -W$$

2. Reversible Adiabatic (ISENTROPIC) Process

Consider a gas contained in the cylinder piston assembly as the system. The cylinder wall and the piston act as adiabatic walls. Suppose the gas is allowed to expand from the initial pressure P_1 to the final pressure P_2 and the opposing pressure is so adjusted that it is equal to inside gas pressure. For such a process, $dW = PdV$

The first-law of thermodynamics will give

$$dU = dQ - dW = -PdV$$

Let us consider the system as an ideal gas which satisfies the relation $Pv = RT$ and $u = u(T)$. Also, we know that $du = c_v dT$.

or

$$du = -Pdv$$

or

$$c_v (T_2 - T_1) = -\int Pdv$$

or

$$c_v (T_2 - T_1) = \int Pdv = w$$

or

For an ideal gas, $h = u + Pv = u + RT$

Therefore,

$$c_p = \left(\frac{\partial h}{\partial T}\right)_p = c_v + R$$

The ratio of specific heats is given by

$$\gamma = \frac{c_p}{c_v} = 1 + \frac{R}{c_v}$$

or

$$c_v = \frac{R}{(\gamma - 1)}$$

Thus,

$$w = c_v(T_1 - T_2) = \frac{R(T_1 - T_2)}{(\gamma - 1)} = \frac{P_1v_1 - P_2v_2}{(\gamma - 1)}$$

Therefore when an ideal gas expands reversibly and adiabatically from the initial state (P_1, v_1, T_1) to final state (P_2, v_2, T_2), the work done per mole of the gas is given by the above expression.

Isentropic Process

In thermodynamics, an isentropic process is an idealized thermodynamic process that is adiabatic and in which the work transfers of the system are frictionless; there is no transfer of matter and the process is reversible. Such an idealized process is useful in engineering as a model of and basis of comparison for real processes.

The word 'isentropic' is occasionally, though not customarily, interpreted in another way, reading it as if its meaning were deducible from its etymology. This is contrary to its original and customarily used definition. In this occasional reading, it means a process in which the entropy of the system remains unchanged, for example because work done on the system includes friction internal to the system, and heat is withdrawn from the system, in just the right amount to compensate for the internal friction, so as to leave the entropy unchanged.

Background

The second law of thermodynamics states that,

$$TdS \geq \delta Q$$

where δQ is the amount of energy the system gains by heating, T is the temperature of the system, and dS is the change in entropy. The equal sign refers to a reversible process, which is an imagined idealized theoretical limit, never actually occurring in physical reality. For an isentropic process, which by definition is reversible, there is no transfer of energy as heat because the process is adiabatic. In an irreversible process of transfer of energy as work, entropy is produced within the system; consequently, in order to maintain constant entropy within the system, energy must be removed from the system as heat during the process.

For reversible processes, an isentropic transformation is carried out by thermally "insulating" the system from its surroundings. Temperature is the thermodynamic conjugate variable to entropy, thus the conjugate process would be an isothermal process in which the system is thermally "connected" to a constant-temperature heat bath.

Isentropic Processes in Thermodynamic Systems

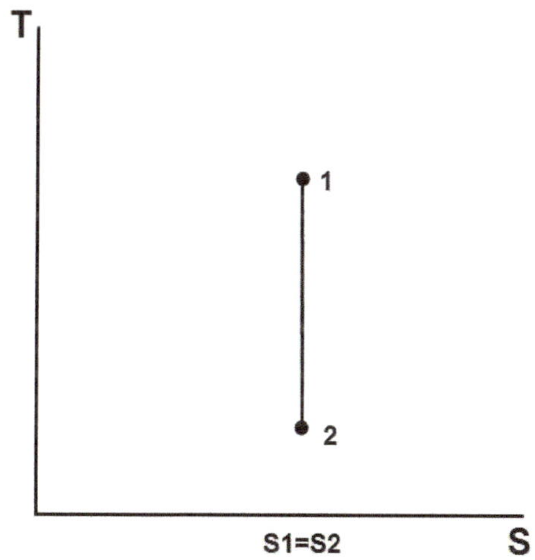

T-s (Entropy vs. Temperature) diagram of an isentropic process, which is a vertical line segment.

The entropy of a given mass does not change during a process that is internally reversible and adiabatic. A process during which the entropy remains constant is called an isentropic process, written $\Delta s = 0$ or $s_1 = s_2$. Some isentropic thermodynamic devices include: pumps, gas compressors, turbines, nozzles, and diffusers.

Isentropic efficiencies of steady-flow devices in Thermodynamic Systems

Most steady-flow devices operate under adiabatic conditions, and the ideal process for these devices is the isentropic process.The parameter that describes how efficiently a device approximates a corresponding isentropic device is called isentropic or adiabatic efficiency.

Isentropic efficiency of Turbines:

$$\eta_T = \frac{\text{Actual Turbine Work}}{\text{Isentropic Turbine Work}} = \frac{W_a}{W_s} \cong \frac{h_1 - h_{2a}}{h_1 - h_{2s}}$$

Isentropic efficiency of Compressors

$$\eta_C = \frac{\text{Isentropic Compressor Work}}{\text{Actual Compressor Work}} = \frac{W_s}{W_a} \cong \frac{h_{2s} - h_1}{h_{2a} - h_1}$$

Isentropic efficiency of Nozzles

$$\eta_N = \frac{\text{Actual KE at Nozzle Exit}}{\text{Isentropic KE at Nozzle Exit}} = \frac{V_{2a}^2}{V_{2s}^2} \cong \frac{h_1 - h_{2a}}{h_1 - h_{2s}}$$

For all the above equations:

h_1 is the specific enthalpy at the entrance state

h_{2a} is the specific enthalpy at the exit state for the actual process

h_{2s} is the specific enthalpy at the exit state for the isentropic process

Isentropic Devices in Thermodynamic Cycles

Ideal Rankine Cycle 1->2 Isentropic compression in a pump

Ideal Rankine Cycle 3->4 Isentropic expansion in a turbine

Ideal Carnot Cycle 2->3 Isentropic expansion

Ideal Carnot Cycle 4->1 Isentropic compression

Ideal Otto Cycle 1->2 Isentropic compression

Ideal Otto Cycle 3->4 Isentropic expansion

Ideal Diesel Cycle 1->2 Isentropic compression

Ideal Diesel Cycle 3->4 Isentropic expansion

Ideal Brayton Cycle 1->2 Isentropic compression in a compressor

Ideal Brayton Cycle 3->4 Isentropic expansion in a turbine

Ideal Vapor-compression refrigeration Cycle 1->2 Isentropic compression in a compressor

NOTE: The isentropic assumptions are only applicable with ideal cycles. Real world cycles have inherent losses due to inefficient compressors and turbines. The real world system are not truly isentropic but are rather idealized as isentropic for calculation purposes.

Isentropic Flow

In fluid dynamics, an isentropic flow is a fluid flow that is both adiabatic and reversible. That is, no heat is added to the flow, and no energy transformations occur due to friction or dissipative effects. For an isentropic flow of a perfect gas, several relations can be derived to define the pressure, density and temperature along a streamline.

Note that energy can be exchanged with the flow in an isentropic transformation, as long as it doesn't happen as heat exchange. An example of such an exchange would be an isentropic expansion or compression that entails work done on or by the flow.

For an isentropic flow, entropy density can vary between different streamlines. If the entropy density is the same everywhere, then the flow is said to be homentropic.

Derivation of the Isentropic Relations

For a closed system, the total change in energy of a system is the sum of the work done and the heat added,

$$dU = \delta W + \delta Q$$

The reversible work done on a system by changing the volume is,

$$dW = -pdV$$

where p is the pressure and V is the volume. The change in enthalpy ($H = U + pV$) is given by,

$$dH = dU + pdV + Vdp$$

Then for a process that is both reversible and adiabatic (i.e. no heat transfer occurs), $\delta Q_{rev} = 0$, and so $dS = \delta Q_{rev} / T = 0$. All reversible adiabatic processes are isentropic. This leads to two important observations,

$dU = \delta W + \delta Q = -pdV + 0$, and

$$dH = \delta W + \delta Q + pdV + Vdp = -pdV + 0 + pdV + Vdp = Vdp$$

Next, a great deal can be computed for isentropic processes of an ideal gas. For any transformation of an ideal gas, it is always true that

$dU = nC_v dT$, and $dH = nC_p dT$.

Using the general results derived above for dU and dH, then

$dU = nC_v dT = -pdV$, and

$dH = nC_p dT = Vdp$.

So for an ideal gas, the heat capacity ratio can be written as,

$$\gamma = \frac{C_p}{C_v} = -\frac{dp/p}{dV/V}$$

For an ideal gas γ is constant. Hence on integrating the above equation, assuming a perfect gas, we get

$pV^{\gamma} = \text{constant}$ i.e.

$$\frac{p_2}{p_1} = \left(\frac{V_1}{V_2}\right)^{\gamma}$$

Using the equation of state for an ideal gas, $pV = nRT$,

$$TV^{\gamma-1} = \text{constant}$$

proof: $PV^{\gamma} = \text{constant} \Rightarrow PVV^{\gamma-1} = \text{constant} \Rightarrow nRTV^{\gamma-1} = \text{constant}$

as nR= constant itself so we can divide both sides by "nR" and still have constant term in the R.H.S.

Ultimate result, $TV^{\gamma-1} = \text{constant}$

$$\frac{p^{\gamma-1}}{T^{\gamma}} = \text{constant}$$

also, for constant $C_p = C_v + R$(per mole),

$\frac{V}{I} = \frac{nR}{p}$ and $p = \frac{nRT}{V}$

$$S_2 - S_1 = nC_p \ln\left(\frac{T_2}{T_1}\right) - nR\ln\left(\frac{p_2}{p_1}\right)$$

$$\frac{S_2 - S_1}{n} = C_p \ln\left(\frac{T_2}{T_1}\right) - R\ln\left(\frac{T_2 V_1}{T_1 V_2}\right) = C_v \ln\left(\frac{T_2}{T_1}\right) + R\ln\left(\frac{V_2}{V_1}\right)$$

Thus for isentropic processes with an ideal gas,

$T_2 = T_1\left(\frac{V_1}{V_2}\right)^{(R/C_v)}$ or $V_2 = V_1\left(\frac{T_1}{T_2}\right)^{(C_v/R)}$

Table of Isentropic Relations for an Ideal Gas

$$\frac{T_2}{T_1} = \left(\frac{p_2}{p_1}\right)^{\frac{\gamma-1}{\gamma}} = \left(\frac{V_1}{V_2}\right)^{(\gamma-1)} = \left(\frac{\rho_2}{\rho_1}\right)^{(\gamma-1)}$$

$$\left(\frac{T_2}{T_1}\right)^{\frac{\gamma}{\gamma-1}} = \frac{p_2}{p_1} = \left(\frac{V_1}{V_2}\right)^{\gamma} = \left(\frac{\rho_1}{\rho_2}\right)^{\gamma}$$

$$\left(\frac{T_1}{T_2}\right)^{\frac{1}{\gamma-1}} = \left(\frac{p_1}{p_2}\right)^{\frac{1}{\gamma}} = \frac{V_2}{V_1} = \frac{\rho_1}{\rho_2}$$

$$\left(\frac{T_2}{T_1}\right)^{\frac{1}{\gamma-1}} = \left(\frac{p_2}{p_1}\right)^{\frac{1}{\gamma}} = \frac{V_1}{V_2} = \frac{\rho_2}{\rho_1}$$

Derived from:

$$pV^{\gamma} = \text{constant}$$

$$pV = mR_sT$$

$$p = \rho R_sT$$

Where:

- p = Pressure
- V = Volume
- γ = Ratio of specific heats = C_p/C_v
- T = Temperature
- m = Mass
- R_s = Gas constant for the specific gas = R/M
- R = Universal gas constant
- M = Molecular weight of the specific gas
- ρ = Density
- C_p = Specific heat at constant pressure
- C_v = Specific heat at constant volume

Polytropic Process

A polytropic process is a thermodynamic process that obeys the relation:

$$pv^n = C$$

where p is the pressure, v is specific volume, n is the polytropic index (a real number), and C is a constant. The polytropic process equation can describe multiple expansion and compression processes which include heat transfer.

If the ideal gas law applies, a process is polytropic if and only if the ratio (K) of energy transfer as heat to energy transfer as work at each infinitesimal step of the process is kept constant:

$$K = \frac{\delta Q}{\delta W} = \text{constant}$$

Particular Cases

Some specific values of n correspond to particular cases:

- $n = 0$ is an isobaric process,
- $n = +\infty$ is an isochoric process.

In addition, when the ideal gas law applies:

- $n = 1$ is an isothermic process,
- $n = \gamma$ is an adiabatic process.

Equivalence between the Polytropic Coefficient and the Ratio of Energy Transfers

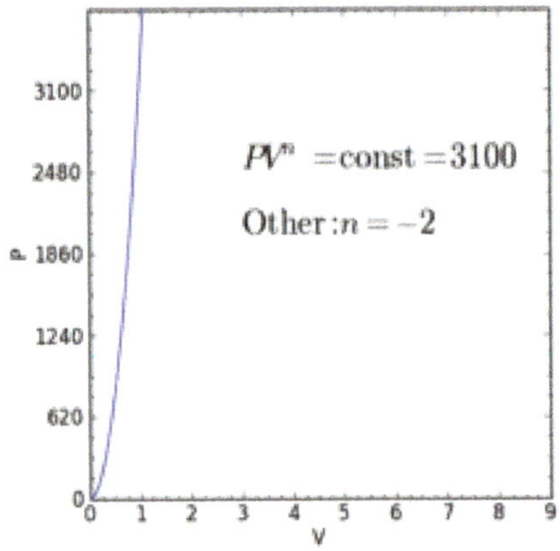

Polytropic processes behave differently with various polytropic indices. Polytropic process can generate other basic thermodynamic processes.

Consider an ideal gas in a closed system undergoing a slow process with negligible changes in kinetic and potential energy. For an infinitesimal step of time, the first law of thermodynamics states that the energy added to a system as heat δq, minus the energy that leaves the system as work δw, is equal to the change in the internal energy du of the system:

$$\delta q - \delta w = du$$

Define the energy transfer ratio,

$$K = \frac{\delta q}{\delta w} \text{ or } \delta q = K\delta w.$$

Transfer of work to the environment can be expressed as $\delta w = pdv$ and internal energy change as $du = nc_v dT$

By substituting the above expressions for δw and δq into the first law:

$$(K-1)pdv = nc_v dT$$

Writing the ideal gas law in differential form

$$pdv + vdp = nRdT \rightarrow \frac{vdp}{pdv} + 1 = \frac{nRdT}{pdv} = \frac{nR(K-1)pdT}{pnc_v dT} = \frac{R(K-1)}{c_v}$$

By Mayer's relation, this becomes:

$$\frac{\upsilon dp}{pdv} = \frac{(c_p - c_v)(K-1)}{c_v} - 1 = (K-1)\gamma - K \rightarrow \frac{dp}{p} + ((1-\gamma)K + \gamma)\frac{dv}{v} = 0$$

where γ is the heat capacity ratio. Assuming K (and γ) remain constant during the transformation, as $\frac{df}{f} = d(\log f)$ this relation can be integrated as

$$d(\log p + ((1-\gamma)K + \gamma)\log v) = 0 \rightarrow pv^{(1-\gamma)K+\gamma} = C$$

where C is a constant.

Thus, the process is polytropic, with the coefficient $n = (1-\gamma)K + \gamma$.

This derivation can be expanded to include polytropic processes in open systems, including instances where the kinetic energy (i.e. Mach number) is significant. It can also be expanded to include irreversible polytropic processes.

Relationship to Ideal Processes

For certain values of the polytropic index, the process will be synonymous with other common processes. Some examples of the effects of varying index values are given in the table.

Variation of polytropic index n		
Polytropic index	**Relation**	**Effects**
$n < 0$	—	Negative exponents reflect a process where work and heat flow simultaneously in or out of the system. In the absence of forces except pressure, such a spontaneous process is not allowed by the second law of thermodynamics; however, negative exponents can be meaningful in some special cases not dominated by thermal interactions, such as in the processes of certain plasmas in astrophysics.

$n = 0$	$p = C$	Equivalent to an isobaric process (constant pressure)
$n = 1$	$pV = C$	Equivalent to an isothermal process (constant temperature), under the assumption of ideal gas law, since then $pV = nRT$.
$1 < n < \gamma$	—	Under the assumption of ideal gas law, heat and work flows go in opposite directions ($K > 0$), such as in vapor compression refrigeration during compression, where the elevated vapour temperature resulting from the work done by the compressor on the vapour leads to some heat loss from the vapour to the cooler surroundings.
$n = \gamma$	—	Equivalent to an isentropic process (adiabatic and reversible, no heat transfer), under the assumption of ideal gas law.
$\gamma < n < \infty$	—	Under the assumption of ideal gas law, heat and work flows go in the same direction ($K < 0$), such as in an internal combustion engine during the power stroke, where heat is lost from the hot combustion products, through the cylinder walls, to the cooler surroundings, at the same time as those hot combustion products push on the piston.
$n = +\infty$	$V = C$	Equivalent to an isochoric process (constant volume)

When the index n is between any two of the former values (0, 1, γ, or ∞), it means that the polytropic curve will cut through (be bounded by) the curves of the two bounding indices.

For an ideal gas, $1 < \gamma < 2$, since by Mayer's relation

$$\gamma = \frac{c_p}{c_v} = \frac{c_v + R}{c_v} = 1 + \frac{R}{c_v} = \frac{c_p}{c_p - R}.$$

Other

A solution to the Lane–Emden equation using a polytropic fluid is known as a polytrope.

Characterisation of Reversible Adiabatic Process

Let us find out the path followed by the system in reaching the final sate starting from the initial state. We have already seen that for an ideal gas

$$c_v dT = -Pdv = -\frac{RT}{v} dv$$

or

$$\frac{dT}{T} = -\frac{R}{c_v}\frac{dv}{v} = -(\gamma - 1)\frac{dv}{v}$$

or

$$1n\frac{T_2}{T_1} = -(\gamma - 1)1n\frac{v_2}{v_1}$$

or

$$\frac{T_2}{T_1} = \left(\frac{v_1}{v_2}\right)^{\gamma-1}$$

or

$$\frac{T_2 v_2^{\gamma-1}}{T_1 v_1^{\gamma-1}} = 1 \qquad or \qquad Tv^{\gamma-1} = \text{constant}$$

Since, $Pv = RT$

$$\frac{T_2}{T_1} = \frac{P_2 v_2}{P_1 v_1}$$

From earlier equations we get

$$\frac{P_2 v_2}{P_1 v_1} = \left(\frac{v_2}{v_1}\right)^{\gamma-1}$$

or

$$\frac{P_2 v_2^{\gamma}}{P_1 v_1^{\gamma}} = 1$$

or

$$Pv^{\gamma} = \text{constant}$$

Polytropic Process

Ideal gas undergoes a reversible-adiabatic process; the path followed by the system is given by

$$Pv^{y} = \text{constant}$$

and the work done per kg of gas

$$w = \frac{P_1 v_1 - P_2 v_2}{y - 1}$$

To generalise,

$$Pv^{n} = \text{constant}; \quad = \text{w} = \frac{P_1 v_1 - P_2 v_2}{y - 1}$$

n is polytropic index (γ is the property of the system, also it indicates reversible-adiabatic process)

$n = 0$, constant pressure process $\quad w = P(v_2 - v_1)$

$n = 1$, constant temperature process $\quad w = q = RT\ln\frac{v_2}{v_1}$

$n = \gamma$, reversible adiabatic process $\quad w = \frac{P_1v_1 - P_2v_2}{\gamma - 1}$

$n = \infty$, constant volume process $\quad w = 0$

Ideal Gas Model

- For many gases, the ideal gas assumption is valid and the $P-v-T$ relationship can be simplified by using the ideal gas equation of state: $Pv = RT$
- For an ideal gas, it has been mentioned that the specific internal energy u is a function of temperature only, $u = u(T)$
- Accordingly, specific enthalpy is a function of temperature only, since $h(T) = u(T) + pv = u(T) + RT$
- Physically, a gas can be considered ideal only when the inter-molar forces are weak.

This usually happens in low pressure and high temperature ranges.

$$c_v = \frac{du}{dT}, c_p = \frac{dh}{dT} = \frac{d(u+Pv)}{dT} = \frac{d(u+RT)}{dT} = c_v + R$$

$$c_p = c_v + R$$

Specific heat ratio $\gamma = c_p/c_v, c_p = c_v + R$

$$c_p(T) = \frac{\gamma R}{\gamma - 1}, c_v(T) = \frac{R}{\gamma - 1}$$

$$du = c_v(T)dT$$

or

$$u(T_2) - u(T_1) = \int_1^2 c_v(T)dT$$

or

$$h(T_2) - h(T_1) = \int_1^2 c_p(T)dT$$

- In some cases, the temperature dependence of the specific heat can be written in polynomial form.
- Otherwise, ideal gas tables are also available. They are easier to use compared to the thermodynamic tables since temperature is the only parameter.

Second Law of Thermodynamics

The second law of thermodynamics states that the total entropy of an isolated system can only increase over time. It can remain constant in ideal cases where the system is in a steady state (equilibrium) or undergoing a reversible process. The increase in entropy accounts for the irreversibility of natural processes, and the asymmetry between future and past.

Historically, the second law was an empirical finding that was accepted as an axiom of thermodynamic theory. Statistical thermodynamics, classical or quantum, explains the microscopic origin of the law.

The second law has been expressed in many ways. Its first formulation is credited to the French scientist Sadi Carnot in 1824, who showed that there is an upper limit to the efficiency of conversion of heat to work in a heat engine.

Introduction

The first law of thermodynamics provides the basic definition of internal energy, associated with all thermodynamic systems, and states the rule of conservation of energy. The second law is concerned with the direction of natural processes. It asserts that a natural process runs only in one sense, and is not reversible. For example, heat always flows spontaneously from hotter to colder bodies, and never the reverse, unless external work is performed on the system. Its modern definition is in terms of entropy.

In a fictive reversible process, an infinitesimal increment in the entropy (dS) of a system is defined to result from an infinitesimal transfer of heat (δQ) to a closed system divided by the common temperature (T) of the system and the surroundings which supply the heat:

$$dS = \frac{\delta Q}{T} \qquad \text{(closed system, idealized fictive reversible process)}$$

Different notations are used for infinitesimal amounts of heat (δ) and infinitesimal amounts of entropy (d) because entropy is a function of state, while heat, like work, is not. For an actually possible infinitesimal process without exchange of matter with the surroundings, the second law requires that the increment in system entropy be greater than that:

$$dS > \frac{\delta Q}{T} \qquad \text{(closed system, actually possible, irreversible process).}$$

This is because a general process for this case may include work being done on the system by its surroundings, which must have frictional or viscous effects inside the system, and because heat transfer actually occurs only irreversibly, driven by a finite temperature difference.

The zeroth law of thermodynamics in its usual short statement allows recognition that two bodies in a relation of thermal equilibrium have the same temperature, especially that a test body has the same temperature as a reference thermometric body. For a body in thermal equilibrium with another, there are indefinitely many empirical temperature scales, in general respectively depending on the properties of a particular reference thermometric body. The second law allows a distinguished temperature scale, which defines an absolute, thermodynamic temperature, independent of the properties of any particular reference thermometric body.

Various Statements of the Law

The second law of thermodynamics may be expressed in many specific ways, the most prominent classical statements being the statement by Rudolf Clausius (1854), the statement by Lord Kelvin (1851), and the statement in axiomatic thermodynamics by Constantin Carathéodory (1909). These statements cast the law in general physical terms citing the impossibility of certain processes. The Clausius and the Kelvin statements have been shown to be equivalent.

Carnot's Principle

The historical origin of the second law of thermodynamics was in Carnot's principle. It refers to a cycle of a Carnot heat engine, fictively operated in the limiting mode of extreme slowness known as quasi-static, so that the heat and work transfers are between subsystems that are always in their own internal states of thermodynamic equilibrium. The Carnot engine is an idealized device of special interest to engineers who are concerned with the efficiency of heat engines. Carnot's principle was recognized by Carnot at a time when the caloric theory of heat was seriously considered, before the recognition of the first law of thermodynamics, and before the mathematical expression of the concept of entropy. Interpreted in the light of the first law, it is physically equivalent to the second law of thermodynamics, and remains valid today. It states

> The efficiency of a quasi-static or reversible Carnot cycle depends only on the temperatures of the two heat reservoirs, and is the same, whatever the working substance. A Carnot engine operated in this way is the most efficient possible heat engine using those two temperatures.

Clausius Statement

The German scientist Rudolf Clausius laid the foundation for the second law of thermodynamics in 1850 by examining the relation between heat transfer and work. His formulation of the second law, which was published in German in 1854, is known as the *Clausius statement*:

Heat can never pass from a colder to a warmer body without some other change, connected therewith, occurring at the same time.

The statement by Clausius uses the concept of 'passage of heat'. As is usual in thermodynamic discussions, this means 'net transfer of energy as heat', and does not refer to contributory transfers one way and the other.

Heat cannot spontaneously flow from cold regions to hot regions without external work being performed on the system, which is evident from ordinary experience of refrigeration, for example. In a refrigerator, heat flows from cold to hot, but only when forced by an external agent, the refrigeration system.

Kelvin Statement

Lord Kelvin expressed the second law as

> It is impossible, by means of inanimate material agency, to derive mechanical effect from any portion of matter by cooling it below the temperature of the coldest of the surrounding objects.

Equivalence of the Clausius and the Kelvin Statements

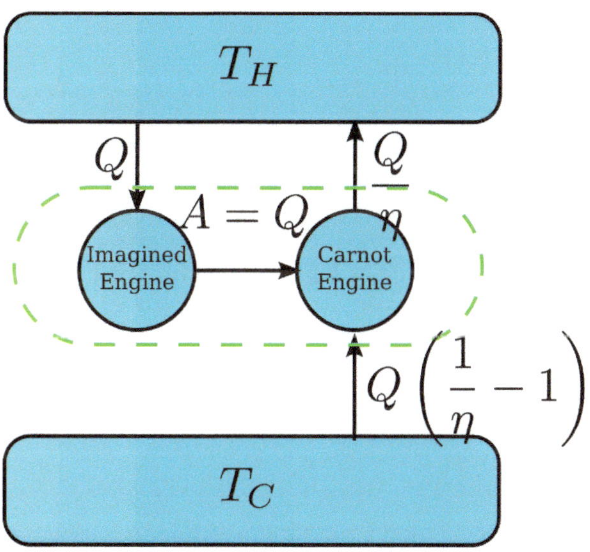

Derive Kelvin Statement from Clausius Statement

Suppose there is an engine violating the Kelvin statement: i.e., one that drains heat and converts it completely into work in a cyclic fashion without any other result. Now pair it with a reversed Carnot engine as shown by the figure. The net and sole effect of this newly created engine consisting of the two engines mentioned is transferring heat $\Delta Q = Q\left(\frac{1}{\eta} - 1\right)$ from the cooler reservoir to the hotter one, which violates the Clausius statement. Thus a violation of the Kelvin statement implies a violation of the Clausius statement, i.e. the Clausius statement implies the Kelvin statement. We can prove in a similar manner that the Kelvin statement implies the Clausius statement, and hence the two are equivalent.

Planck's Proposition

Planck offered the following proposition as derived directly from experience. This is sometimes regarded as his statement of the second law, but he regarded it as a starting point for the derivation of the second law.

> *It is impossible to construct an engine which will work in a complete cycle, and produce no effect except the raising of a weight and cooling of a heat reservoir.*

Relation between Kelvin's Statement and Planck's Proposition

It is almost customary in textbooks to speak of the "Kelvin-Planck statement" of the law, as for example in the text by ter Haar and Wergeland. One text gives a statement very like Planck's proposition, but attributes it to Kelvin without mention of Planck. One monograph quotes Planck's proposition as the "Kelvin-Planck" formulation, the text naming Kelvin as its author, though it correctly cites Planck in its references. The reader may compare the two statements quoted just above here.

Planck's Statement

Planck stated the second law as follows.

> *Every process occurring in nature proceeds in the sense in which the sum of the entropies of all bodies taking part in the process is increased. In the limit, i.e. for reversible processes, the sum of the entropies remains unchanged.*

Rather like Planck's statement is that of Uhlenbeck and Ford for *irreversible phenomena*.

> ... in an irreversible or spontaneous change from one equilibrium state to another (as for example the equalization of temperature of two bodies A and B, when brought in contact) the entropy always increases.

Principle of Carathéodory

Constantin Carathéodory formulated thermodynamics on a purely mathematical axiomatic foundation. His statement of the second law is known as the Principle of Carathéodory, which may be formulated as follows:

In every neighborhood of any state S of an adiabatically enclosed system there are states inaccessible from S.

With this formulation, he described the concept of adiabatic accessibility for the first time and provided the foundation for a new subfield of classical thermodynamics, often called geometrical thermodynamics. It follows from Carathéodory's principle that quantity of energy quasi-statically transferred as heat is a holonomic process function, in other words, $\delta Q = TdS$.

Though it is almost customary in textbooks to say that Carathéodory's principle expresses the second law and to treat it as equivalent to the Clausius or to the Kelvin-Planck statements, such is not the case. To get all the content of the second law, Carathéodory's principle needs to be supplemented by Planck's principle, that isochoric work always increases the internal energy of a closed system that was initially in its own internal thermodynamic equilibrium.

Planck's Principle

In 1926, Max Planck wrote an important paper on the basics of thermodynamics. He indicated the principle

> The internal energy of a closed system is increased by an adiabatic process, throughout the duration of which, the volume of the system remains constant.

This formulation does not mention heat and does not mention temperature, nor even entropy, and does not necessarily implicitly rely on those concepts, but it implies the content of the second law. A closely related statement is that "Frictional pressure never does positive work." Using a now-obsolete form of words, Planck himself wrote: "The production of heat by friction is irreversible."

Not mentioning entropy, this principle of Planck is stated in physical terms. It is very closely related to the Kelvin statement given just above. It is relevant that for a system at constant volume and mole numbers, the entropy is a monotonic function of the internal energy. Nevertheless, this principle of Planck is not actually Planck's preferred statement of the second law, which is quoted above, in a previous sub-section of the present section of this present article, and relies on the concept of entropy.

A statement that in a sense is complementary to Planck's principle is made by Borgnakke and Sonntag. They do not offer it as a full statement of the second law:

> ...there is only one way in which the entropy of a [closed] system can be decreased, and that is to transfer heat from the system.

Differing from Planck's just foregoing principle, this one is explicitly in terms of entropy change. Of course, removal of matter from a system can also decrease its entropy.

Statement for a System that has a Known Expression of its Internal Energy as a Function of its Extensive State Variables

The second law has been shown to be equivalent to the internal energy U being a weakly convex function, when written as a function of extensive properties (mass, volume, entropy, ...).

Corollaries

Perpetual Motion of the Second Kind

Before the establishment of the Second Law, many people who were interested in inventing a perpetual motion machine had tried to circumvent the restrictions of first law of thermodynamics by extracting the massive internal energy of the environment as the power of the machine. Such a machine is called a "perpetual motion machine of the second kind". The second law declared the impossibility of such machines.

Carnot Theorem

Carnot's theorem (1824) is a principle that limits the maximum efficiency for any possible engine. The efficiency solely depends on the temperature difference between the hot and cold thermal reservoirs. Carnot's theorem states:

- All irreversible heat engines between two heat reservoirs are less efficient than a Carnot engine operating between the same reservoirs.
- All reversible heat engines between two heat reservoirs are equally efficient with a Carnot engine operating between the same reservoirs.

In his ideal model, the heat of caloric converted into work could be reinstated by reversing the motion of the cycle, a concept subsequently known as thermodynamic reversibility. Carnot, however, further postulated that some caloric is lost, not being converted to mechanical work. Hence, no real heat engine could realise the Carnot cycle's reversibility and was condemned to be less efficient.

Though formulated in terms of caloric, rather than entropy, this was an early insight into the second law.

Clausius Inequality

The Clausius theorem (1854) states that in a cyclic process

$$\oint \frac{\delta Q}{T} \leq 0.$$

The equality holds in the reversible case and the '<' is in the irreversible case. The reversible case is used to introduce the state function entropy. This is because in cyclic processes the variation of a state function is zero from state functionality.

Thermodynamic Temperature

For an arbitrary heat engine, the efficiency is:

$$\eta = \frac{W_n}{q_H} = \frac{q_H - q_C}{q_H} = 1 - \frac{q_C}{q_H}$$

where W_n is for the net work done per cycle. Thus the efficiency depends only on q_C/q_H.

Carnot's theorem states that all reversible engines operating between the same heat reservoirs are equally efficient. Thus, any reversible heat engine operating between temperatures T_1 and T_2 must have the same efficiency, that is to say, the efficiency is the function of temperatures only: $\frac{q_C}{q_H} = f(T_H, T_C)$

In addition, a reversible heat engine operating between temperatures T_1 and T_3 must have the same efficiency as one consisting of two cycles, one between T_1 and another (intermediate) temperature T_2, and the second between T_2 and T_3. This can only be the case if

$$f(T_1,T_3) = \frac{q_3}{q_1} = \frac{q_2 q_3}{q_1 q_2} = f(T_1,T_2) f(T_2,T_3).$$

Now consider the case where T_1 is a fixed reference temperature: the temperature of the triple point of water. Then for any T_2 and T_3,

$$f(T_2,T_3) = \frac{f(T_1,T_3)}{f(T_1,T_2)} = \frac{273.16 \cdot f(T_1,T_3)}{273.16 \cdot f(T_1,T_2)}.$$

Therefore, if thermodynamic temperature is defined by

$$T = 273.16 \cdot f(T_1,T)$$

then the function f, viewed as a function of thermodynamic temperature, is simply

$$f(T_2,T_3) = \frac{T_3}{T_2},$$

and the reference temperature T_1 will have the value 273.16. (Of course any reference temperature and any positive numerical value could be used—the choice here corresponds to the Kelvin scale.)

Entropy

According to the Clausius equality, for a reversible process

$$\oint \frac{\delta Q}{T} = 0$$

That means the line integral $\int_L \frac{\delta Q}{T}$ is path independent.

So we can define a state function S called entropy, which satisfies

$$dS = \frac{\delta Q}{T}$$

With this we can only obtain the difference of entropy by integrating the above formula. To obtain the absolute value, we need the Third Law of Thermodynamics, which states that S=0 at absolute zero for perfect crystals.

For any irreversible process, since entropy is a state function, we can always connect the initial and terminal states with an imaginary reversible process and integrating on that path to calculate the difference in entropy.

Now reverse the reversible process and combine it with the said irreversible process. Applying Clausius inequality on this loop,

$$-\Delta S + \int \frac{\delta Q}{T} = \oint \frac{\delta Q}{T} < 0$$

Thus,

$$\Delta S \geq \int \frac{\delta Q}{T}$$

where the equality holds if the transformation is reversible.

Notice that if the process is an adiabatic process, then $\delta Q = 0$, so $\Delta S \geq 0$.

Energy, Available Useful Work

An important and revealing idealized special case is to consider applying the Second Law to the scenario of an isolated system (called the total system or universe), made up of two parts: a sub-system of interest, and the sub-system's surroundings. These surroundings are imagined to be so large that they can be considered as an *unlimited* heat reservoir at temperature T_R and pressure P_R — so that no matter how much heat is transferred to (or from) the sub-system, the temperature of the surroundings will remain T_R; and no matter how much the volume of the sub-system expands (or contracts), the pressure of the surroundings will remain P_R.

Whatever changes to dS and dS_R occur in the entropies of the sub-system and the surroundings individually, according to the Second Law the entropy S_{tot} of the isolated total system must not decrease:

$$dS_{\text{tot}} = dS + dS_R \geq 0$$

According to the First Law of Thermodynamics, the change dU in the internal energy of the sub-system is the sum of the heat δq added to the sub-system, *less* any work δw done *by* the sub-system, *plus* any net chemical energy entering the sub-system $d\sum \mu_{iR} N_i$, so that:

$$dU = \delta q - \delta w + d\left(\sum \mu_{iR} N_i\right)$$

where μ_{iR} are the chemical potentials of chemical species in the external surroundings.

Now the heat leaving the reservoir and entering the sub-system is

$$\delta q = T_R(-dS_R) \le T_R dS$$

where we have first used the definition of entropy in classical thermodynamics (alternatively, in statistical thermodynamics, the relation between entropy change, temperature and absorbed heat can be derived); and then the Second Law inequality from above.

It therefore follows that any net work δw done by the sub-system must obey

$$\delta w \le -dU + T_R dS + \sum \mu_{iR} dN_i$$

It is useful to separate the work δw done by the subsystem into the *useful* work δw_u that can be done *by* the sub-system, over and beyond the work $p_R\ dV$ done merely by the sub-system expanding against the surrounding external pressure, giving the following relation for the useful work (exergy) that can be done:

$$\delta w_u \le -d(U - T_R S + p_R V - \sum \mu_{iR} N_i)$$

It is convenient to define the right-hand-side as the exact derivative of a thermodynamic potential, called the *availability* or *exergy E* of the subsystem,

$$E = U - T_R S + p_R V - \sum \mu_{iR} N_i$$

The Second Law therefore implies that for any process which can be considered as divided simply into a subsystem, and an unlimited temperature and pressure reservoir with which it is in contact,

$$dE + \delta w_u \le 0$$

i.e. the change in the subsystem's exergy plus the useful work done *by* the subsystem (or, the change in the subsystem's exergy less any work, additional to that done by the pressure reservoir, done *on* the system) must be less than or equal to zero.

In sum, if a proper *infinite-reservoir-like* reference state is chosen as the system surroundings in the real world, then the Second Law predicts a decrease in *E* for an irreversible process and no change for a reversible process.

$dS_{tot} \ge 0$ Is equivalent to $dE + \delta w_u \le 0$

This expression together with the associated reference state permits a design engineer working at the macroscopic scale (above the thermodynamic limit) to utilize the Second Law without directly measuring or considering entropy change in a total isolated system. Those changes have already been considered by the assumption that the system under consideration can reach equilibrium with the reference state without altering the reference state. An efficiency for a process or collection of processes that compares it to the reversible ideal may also be found.

This approach to the Second Law is widely utilized in engineering practice, environmental accounting, systems ecology, and other disciplines.

History

Nicolas Léonard Sadi Carnot in the traditional uniform of a student of the École Polytechnique.

The first theory of the conversion of heat into mechanical work is due to Nicolas Léonard Sadi Carnot in 1824. He was the first to realize correctly that the efficiency of this conversion depends on the difference of temperature between an engine and its environment.

Recognizing the significance of James Prescott Joule's work on the conservation of energy, Rudolf Clausius was the first to formulate the second law during 1850, in this form: heat does not flow *spontaneously* from cold to hot bodies. While common knowledge now, this was contrary to the caloric theory of heat popular at the time, which considered heat as a fluid. From there he was able to infer the principle of Sadi Carnot and the definition of entropy (1865).

Established during the 19th century, the Kelvin-Planck statement of the Second Law says, "It is impossible for any device that operates on a cycle to receive heat from a single reservoir and produce a net amount of work." This was shown to be equivalent to the statement of Clausius.

The ergodic hypothesis is also important for the Boltzmann approach. It says that, over long periods of time, the time spent in some region of the phase space of microstates with the same energy is proportional to the volume of this region, i.e. that all accessible microstates are equally probable over a long period of time. Equivalently, it says that time average and average over the statistical ensemble are the same.

There is a traditional doctrine, starting with Clausius, that entropy can be understood in terms of molecular 'disorder' within a macroscopic system. This doctrine is obsolescent.

Account Given by Clausius

Rudolf Clausius

In 1856, the German physicist Rudolf Clausius stated what he called the "second fundamental theorem in the mechanical theory of heat" in the following form:

$$\int \frac{\delta Q}{T} = -N$$

where Q is heat, T is temperature and N is the "equivalence-value" of all uncompensated transformations involved in a cyclical process. Later, in 1865, Clausius would come to define "equivalence-value" as entropy. On the heels of this definition, that same year, the most famous version of the second law was read in a presentation at the Philosophical Society of Zurich on April 24, in which, in the end of his presentation, Clausius concludes:

The entropy of the universe tends to a maximum.

This statement is the best-known phrasing of the second law. Because of the looseness of its language, e.g. universe, as well as lack of specific conditions, e.g. open, closed, or isolated, many people take this simple statement to mean that the second law of thermodynamics applies virtually to every subject imaginable. This, of course, is not true; this statement is only a simplified version of a more extended and precise description.

In terms of time variation, the mathematical statement of the second law for an isolated system undergoing an arbitrary transformation is:

$$\frac{dS}{dt} \geq 0$$

where

S is the entropy of the system and

t is time.

The equality sign applies after equilibration. An alternative way of formulating of the second law for isolated systems is:

$$\frac{dS}{dt} = \dot{S}_i \text{ with } \dot{S}_i \geq 0$$

with $\dot{S}_i$ the sum of the rate of entropy production by all processes inside the system. The advantage of this formulation is that it shows the effect of the entropy production. The rate of entropy production is a very important concept since it determines (limits) the efficiency of thermal machines. Multiplied with ambient temperature T_a it gives the so-called dissipated energy $P_{diss} = T_a \dot{S}_i$.

The expression of the second law for closed systems (so, allowing heat exchange and moving boundaries, but not exchange of matter) is:

$$\frac{dS}{dt} = \frac{\dot{Q}}{T} + \dot{S}_i \text{ with } \dot{S}_i \geq 0$$

Here

$\dot{Q}$ is the heat flow into the system

T is the temperature at the point where the heat enters the system.

The equality sign holds in the case that only reversible processes take place inside the system. If irreversible processes take place (which is the case in real systems in operation) the >-sign holds. If heat is supplied to the system at several places we have to take the algebraic sum of the corresponding terms.

For open systems (also allowing exchange of matter):

$$\frac{dS}{dt} = \frac{\dot{Q}}{T} + \dot{S} + \dot{S}_i \text{ with } \dot{S}_i \geq 0$$

Here $\dot{S}$ is the flow of entropy into the system associated with the flow of matter entering the system. It should not be confused with the time derivative of the entropy. If matter is supplied at several places we have to take the algebraic sum of these contributions.

Statistical Mechanics

Statistical mechanics gives an explanation for the second law by postulating that a material is composed of atoms and molecules which are in constant motion. A particular set of positions and velocities for each particle in the system is called a microstate of the system and because of the constant motion, the system is constantly changing its microstate. Statistical mechanics postulates that, in equilibrium, each microstate that the system might be in is equally likely to occur, and when this assumption is made, it leads directly to the conclusion that the second law must hold in a statistical sense. That is, the second law will hold on average, with a statistical variation on the order of $1/\sqrt{N}$ where N is the number of particles in the system. For everyday (macroscopic) situations, the probability that the second law will be violated is practically zero. However, for systems with a small number of particles, thermodynamic parameters, including the entropy, may show significant statistical deviations from that predicted by the second law. Classical thermodynamic theory does not deal with these statistical variations.

Derivation from Statistical Mechanics

The first mechanical argument of the Kinetic theory of gases that molecular collisions entail an equalization of temperatures and hence a tendency towards equilibrium was due to James Clerk Maxwell in 1860; Ludwig Boltzmann with his H-theorem of 1872 also argued that due to collisions gases should over time tend toward the Maxwell-Boltzmann distribution.

Due to Loschmidt's paradox, derivations of the Second Law have to make an assumption regarding the past, namely that the system is uncorrelated at some time in the past; this allows for simple probabilistic treatment. This assumption is usually thought as a boundary condition, and thus the second Law is ultimately a consequence of the initial conditions somewhere in the past, probably at the beginning of the universe (the Big Bang), though other scenarios have also been suggested.

Given these assumptions, in statistical mechanics, the Second Law is not a postulate, rather it is a consequence of the fundamental postulate, also known as the equal prior probability postulate, so long as one is clear that simple probability arguments are applied only to the future, while for the past there are auxiliary sources of information which tell us that it was low entropy. The first part of the second law, which states that the entropy of a thermally isolated system can only increase, is a trivial consequence of the equal prior probability postulate, if we restrict the notion of the entropy to systems in thermal equilibrium. The entropy of an isolated system in thermal equilibrium containing an amount of energy of E is:

$$S = k_B \ln\left[\Omega(E)\right]$$

where $\Omega(E)$ is the number of quantum states in a small interval between E and $E+\delta E$. Here δE is a macroscopically small energy interval that is kept fixed. Strictly speaking this means that the entropy depends on the choice of δE. However, in the thermodynamic limit (i.e. in the limit of infinitely large system size), the specific entropy (entropy per unit volume or per unit mass) does not depend on δE.

Suppose we have an isolated system whose macroscopic state is specified by a number of variables. These macroscopic variables can, e.g., refer to the total volume, the positions of pistons in the system, etc. Then Ω will depend on the values of these variables. If a variable is not fixed, (e.g. we do not clamp a piston in a certain position), then because all the accessible states are equally likely in equilibrium, the free variable in equilibrium will be such that Ω is maximized as that is the most probable situation in equilibrium.

If the variable was initially fixed to some value then upon release and when the new equilibrium has been reached, the fact the variable will adjust itself so that Ω is maximized, implies that the entropy will have increased or it will have stayed the same (if the value at which the variable was fixed happened to be the equilibrium value). Suppose we start from an equilibrium situation and we suddenly remove a constraint on a variable. Then right after we do this, there are a number Ω of accessible microstates, but equilibrium has not yet been reached, so the actual probabilities of the system being in some accessible state are not yet equal to the prior probability of $1/\Omega$. . We have already seen that in the final equilibrium state, the entropy will have increased or have stayed the same relative to the previous equilibrium state. Boltzmann's H-theorem, however, proves that the quantity H increases monotonically as a function of time during the intermediate out of equilibrium state.

Derivation of the Entropy Change for Reversible Processes

The second part of the Second Law states that the entropy change of a system undergoing a reversible process is given by:

$$dS = \frac{\delta Q}{T}$$

where the temperature is defined as:

$$\frac{1}{k_B T} \equiv \beta \equiv \frac{d\ln\left[\Omega(E)\right]}{dE}$$

See here for the justification for this definition. Suppose that the system has some external parameter, x, that can be changed. In general, the energy eigenstates of the system will depend on x. According to the adiabatic theorem of quantum mechanics, in the limit of an infinitely slow change of the system's Hamiltonian, the system will stay in the same energy eigenstate and thus change its energy according to the change in energy of the energy eigenstate it is in.

The generalized force, X, corresponding to the external variable x is defined such that Xdx is the work performed by the system if x is increased by an amount dx. E.g., if x is the volume, then X is the pressure. The generalized force for a system known to be in energy eigenstate E_r is given by:

$$X = -\frac{dE_r}{dx}$$

Since the system can be in any energy eigenstate within an interval of δE, we define the generalized force for the system as the expectation value of the above expression:

$$X = -\left\langle \frac{dE_r}{dx} \right\rangle$$

To evaluate the average, we partition the $\Omega(E)$ energy eigenstates by counting how many of them have a value for $\frac{dE_r}{dx}$ within a range between Y and $Y + \delta Y$. Calling this number $\Omega_Y(E)$, we have:

$$\Omega(E) = \sum_Y \Omega_Y(E)$$

The average defining the generalized force can now be written:

$$X = -\frac{1}{\Omega(E)} \sum_Y Y\Omega_Y(E)$$

We can relate this to the derivative of the entropy with respect to x at constant energy E as follows. Suppose we change x to x + dx. Then $\Omega(E)$ will change because the energy eigenstates depend on x, causing energy eigenstates to move into or out of the range between E and $E + \delta E$. Let's focus again on the energy eigenstates for which $\frac{dE_r}{dx}$ lies within the range between Y and $Y + \delta Y$. Since these energy eigenstates increase in energy by Y dx, all such energy eigenstates that are in the interval ranging from E – Y dx to E move from below E to above E. There are

$$N_Y(E) - \frac{\Omega_Y(E)}{\delta E} Ydx$$

such energy eigenstates. If $Ydx \leq \delta E$, all these energy eigenstates will move into the range between E and $E + \delta E$ and contribute to an increase in Ω. The number of energy eigenstates that move from below $E + \delta E$ to above $E + \delta E$ is, of course, given by $N_Y(E + \delta E)$.. The difference

$$N_Y(E) - N_Y(E + \delta E)$$

is thus the net contribution to the increase in Ω. Note that if Y dx is larger than δE there will be the energy eigenstates that move from below E to above $E + \delta E$. They are counted in both $N_Y(E)$ and $N_Y(E + \delta E)$, therefore the above expression is also valid in that case.

Expressing the above expression as a derivative with respect to E and summing over Y yields the expression:

$$\left(\frac{\partial \Omega}{\partial x}\right)_E = -\sum_Y Y\left(\frac{\partial \Omega_Y}{\partial E}\right)_x = \left(\frac{\partial (\Omega X)}{\partial E}\right)_x$$

The logarithmic derivative of Ω with respect to x is thus given by:

$$\left(\frac{\partial \ln(\Omega)}{\partial x}\right)_E = \beta X + \left(\frac{\partial X}{\partial E}\right)_x$$

The first term is intensive, i.e. it does not scale with system size. In contrast, the last term scales as the inverse system size and will thus vanishes in the thermodynamic limit. We have thus found that:

$$\left(\frac{\partial S}{\partial x}\right)_E = \frac{X}{T}$$

Combining this with

$$\left(\frac{\partial S}{\partial E}\right)_x = \frac{1}{T}$$

Gives:

$$dS = \left(\frac{\partial S}{\partial E}\right)_x dE + \left(\frac{\partial S}{\partial x}\right)_E dx = \frac{dE}{T} + \frac{X}{T}dx = \frac{\delta Q}{T}$$

Derivation for Systems Described by the Canonical Ensemble

If a system is in thermal contact with a heat bath at some temperature T then, in equilibrium, the probability distribution over the energy eigenvalues are given by the canonical ensemble:

$$P_j = \frac{\exp\left(-\frac{E_j}{k_B T}\right)}{Z}$$

Here Z is a factor that normalizes the sum of all the probabilities to 1, this function is known as the partition function. We now consider an infinitesimal reversible change in the temperature and in the external parameters on which the energy levels depend. It follows from the general formula for the entropy:

$$S = -k_B \sum_j P_j \ln(P_j)$$

that

$$dS = -k_B \sum_j \ln(P_j) dP_j$$

Inserting the formula for P_j for the canonical ensemble in here gives:

$$dS = \frac{1}{T}\sum_j E_j dP_j = \frac{1}{T}\sum_j d(E_j P_j) - \frac{1}{T}\sum_j P_j dE_j = \frac{dE + \delta W}{T} = \frac{\delta Q}{T}$$

Living Organisms

There are two principal ways of formulating thermodynamics, (a) through passages from one state of thermodynamic equilibrium to another, and (b) through cyclic processes, by which the system is left unchanged, while the total entropy of the surroundings is increased. These two ways help to understand the processes of life. This topic is mostly beyond the scope of this present article, but has been considered by several authors, such as Erwin Schrödinger, Léon Brillouin and Isaac Asimov. It is also the topic of current research.

To a fair approximation, living organisms may be considered as examples of (b). Approximately, an animal's physical state cycles by the day, leaving the animal nearly unchanged. Animals take in food, water, and oxygen, and, as a result of metabolism, give out breakdown products and heat. Plants take in radiative energy from the sun, which may be regarded as heat, and carbon dioxide and water. They give out oxygen. In this way they grow. Eventually they die, and their remains rot. This can be regarded as a cyclic process. Overall, the sunlight is from a high temperature source, the sun, and its energy is passed to a lower temperature sink, the soil. This is an increase of entropy of the surroundings of the plant. Thus animals and plants obey the second law of thermodynamics, considered in terms of cyclic processes. Simple concepts of efficiency of heat engines are hardly applicable to this problem because they assume closed systems.

From the thermodynamic viewpoint that considers (a), passages from one equilibrium state to another, only a roughly approximate picture appears, because living organisms are never in states of thermodynamic equilibrium. Living organisms must often be considered as open systems, because they take in nutrients and give out waste products. Thermodynamics of open systems is currently often considered in terms of passages from one state of thermodynamic equilibrium to another, or in terms of flows in the approximation of local thermodynamic equilibrium. The problem for living organisms may be further simplified by the approximation of assuming a steady state with unchanging flows. General principles of entropy production for such approximations are subject to unsettled current debate or research. Nevertheless, ideas derived from this viewpoint on the second law of thermodynamics are enlightening about living creatures.

Gravitational Systems

In systems that do not require for their descriptions the general theory of relativity, bodies always have positive heat capacity, meaning that the temperature rises with energy. Therefore, when energy flows from a high-temperature object to a low-temperature object, the source temperature is decreased while the sink temperature is increased; hence temperature differences tend to diminish over time. This is not always the case for systems in which the gravitational force is important and the general theory of relativity is required. Such systems can spontaneously change towards uneven spread of mass and energy. This applies to the universe in large scale, and consequently it may be difficult or impossible to apply the second law to it. Beyond this, the thermodynamics of systems described by the general theory of relativity is beyond the scope of the present article.

Non-equilibrium States

The theory of classical or equilibrium thermodynamics is idealized. A main postulate or assumption, often not even explicitly stated, is the existence of systems in their own internal states of thermodynamic equilibrium. In general, a region of space containing a physical system at a given time, that may be found in nature, is not in thermodynamic equilibrium, read in the most stringent terms. In looser terms, nothing in the entire universe is or has ever been truly in exact thermodynamic equilibrium.

For purposes of physical analysis, it is often enough convenient to make an assumption of thermodynamic equilibrium. Such an assumption may rely on trial and error for its justification. If the assumption is justified, it can often be very valuable and useful because it makes available the theory of thermodynamics. Elements of the equilibrium assumption are that a system is observed to be unchanging over an indefinitely long time, and that there are so many particles in a system, that its particulate nature can be entirely ignored. Under such an equilibrium assumption, in general, there are no macroscopically detectable fluctuations. There is an exception, the case of critical states, which exhibit to the naked eye the phenomenon of critical opalescence. For laboratory studies of critical states, exceptionally long observation times are needed.

In all cases, the assumption of thermodynamic equilibrium, once made, implies as a consequence that no putative candidate "fluctuation" alters the entropy of the system.

It can easily happen that a physical system exhibits internal macroscopic changes that are fast enough to invalidate the assumption of the constancy of the entropy. Or that a physical system has so few particles that the particulate nature is manifest in observable fluctuations. Then the assumption of thermodynamic equilibrium is to be abandoned. There is no unqualified general definition of entropy for non-equilibrium states.

There are intermediate cases, in which the assumption of local thermodynamic equilibrium is a very good approximation, but strictly speaking it is still an approximation, not theoretically ideal. For non-equilibrium situations in general, it may be useful to consider statistical mechanical definitions of other quantities that may be conveniently called 'entropy', but they should not be confused or conflated with thermodynamic entropy properly defined for the second law. These other quantities indeed belong to statistical mechanics, not to thermodynamics, the primary realm of the second law.

The physics of macroscopically observable fluctuations is beyond the scope of this article.

Arrow of Time

The second law of thermodynamics is a physical law that is not symmetric to reversal of the time direction.

The second law has been proposed to supply an explanation of the difference between moving forward and backwards in time, such as why the cause precedes the effect (the causal arrow of time).

Irreversibility

Irreversibility in thermodynamic processes is a consequence of the asymmetric character of thermodynamic operations, and not of any internally irreversible microscopic properties of the bodies. Thermodynamic operations are macroscopic external interventions imposed on the participating bodies, not derived from their internal properties. There are reputed "paradoxes" that arise from failure to recognize this.

Loschmidt's Paradox

Loschmidt's paradox, also known as the reversibility paradox, is the objection that it should not be possible to deduce an irreversible process from the time-symmetric dynamics that describe the microscopic evolution of a macroscopic system.

In the opinion of Schrödinger, "It is now quite obvious in what manner you have to reformulate the law of entropy—or for that matter, all other irreversible statements—so that they be capable of being derived from reversible models. You must not speak of one isolated system but at least of two, which you may for the moment consider isolated from the rest of the world, but not always from each other." The two systems are isolated from each other by the wall, until it is removed by the thermodynamic operation, as envisaged by the law. The thermodynamic operation is externally imposed, not subject to the reversible microscopic dynamical laws that govern the constituents of the systems. It is the cause of the irreversibility. The statement of the law in this present article complies with Schrödinger's advice. The cause–effect relation is logically prior to the second law, not derived from it.

Poincaré Recurrence Theorem

The Poincaré recurrence theorem considers a theoretical microscopic description of an isolated physical system. This may be considered as a model of a thermodynamic system after a thermodynamic operation has removed an internal wall. The system will, after a sufficiently long time, return to a microscopically defined state very close to the initial one. The Poincaré recurrence time is the length of time elapsed until the return. It is exceedingly long, likely longer than the life of the universe, and depends sensitively on the geometry of the wall that was removed by the thermodynamic operation. The recurrence theorem may be perceived as apparently contradicting the second law of thermodynamics. More obviously, however, it is simply a microscopic model of thermodynamic equilibrium in an isolated system formed by removal of a wall between two systems. For a typical thermodynamical system, the recurrence time is so large (many many times longer than the lifetime of the universe) that, for all practical purposes, one cannot observe the recurrence. One might wish, nevertheless, to imagine that one could wait for the Poincaré recurrence, and then re-insert the wall that was removed by the thermodynamic operation. It is then evident that the appearance of irreversibility is due to the utter unpredictability of the Poincaré recurrence given only that the initial state was one of thermodynamic equilibrium, as is the case in macroscopic thermodynamics. Even if one could wait for it, one has no practical possibility of picking the right instant at which to re-insert the wall. The Poincaré recurrence theorem provides a solution to Loschmidt's paradox. If an isolated thermodynamic system could be monitored over increasingly many multiples of the average Poincaré recurrence time, the thermodynamic behavior of the system would become invariant under time reversal.

James Clerk Maxwell

Maxwell's Demon

James Clerk Maxwell imagined one container divided into two parts, *A* and *B*. Both parts are filled with the same gas at equal temperatures and placed next to each other, separated by a wall. Observing the molecules on both sides, an imaginary demon guards a microscopic trapdoor in the wall. When a faster-than-average molecule from *A* flies towards the trapdoor, the demon opens it, and the molecule will fly from *A* to *B*. The average speed of the molecules in *B* will have increased while in *A* they will have slowed down on average. Since average molecular speed corresponds to temperature, the temperature decreases in *A* and increases in *B*, contrary to the second law of thermodynamics.

One response to this question was suggested in 1929 by Leó Szilárd and later by Léon Brillouin. Szilárd pointed out that a real-life Maxwell's demon would need to have some means of measuring molecular speed, and that the act of acquiring information would require an expenditure of energy.

Maxwell's demon repeatedly alters the permeability of the wall between A and B. It is therefore performing thermodynamic operations on a microscopic scale, not just observing ordinary spontaneous or natural macroscopic thermodynamic processes.

Reversibility, Irreversibility and Carnot Cycle

The second law of thermodynamics distinguishes between reversible and irreversible processes.

- If a process can proceed in either direction without violating the second law of thermodynamics, it is reversible process. A reversible process is carried out infinitely slowly with an infinitesimal gradient, so that every state passed through by the system is an equilibrium state. So, a reversible process is a quasi-static process which can proceed in either direction.
- Given a process, if the attempt to reverse its direction leads to a violation of the second law of thermodynamics, then the given process is irreversible.

Any natural process carried out with a finite gradient is an irreversible process . A reversible process which consists of a succession of equilibrium states, is an idealized hypothetical process, approached only as a limit. It is said to be an asymptote to reality, All spontaneous processes are irreversible.

Irreversible Processes

The example of irreversible processes are: Motion with friction, free expansion, Expansion/ compression with finite pressure difference, Energy transfer as heat with finite ΔT , Mixing of matter at different states, Mixing of non-identical gases.

Reversible Processes

The processes which can be idealized as reversible are: Motion without friction, Expansion/compression with infinitesimal pressure difference, Energy transfer as heat with infinitesimal temperature difference.

Carnot Cycle

A French engineer Sadi Carnot was the first to introduce the idea of reversible cycle. From the second law, it has been observed that the efficiency of a heat engine is less than unity. If the efficiency of heat engine is less than unity, what is the maximum efficiency of a heat engine? This can be answered by considering the Carnot cycle. The concept of carnot cycle is executed via Carnot engine.

Carnot Engine

Let us consider the operation of a hypothetical engine which employs the Carnot cycle. The Carnot engine consists of a cylinder-piston assembly in which a certain amount of gas(working fluid) is enclosed. Refer to Figure representing the Carnot cycle.

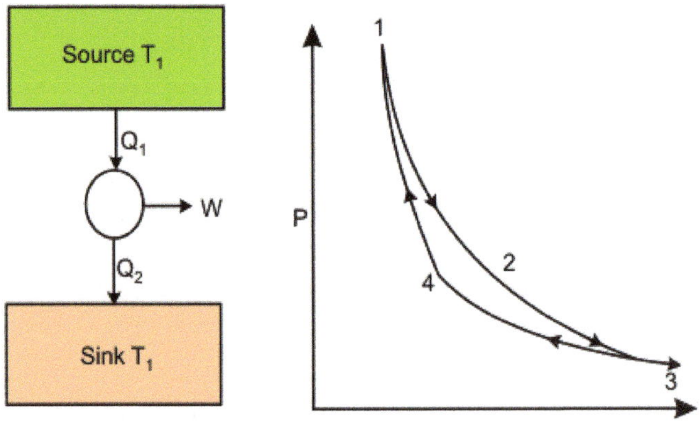

Reversible Isothermal Heat Addition

In the first process, the cylinder head is brought into contact with a source at temperature T_1. The gas inside the cylinder is also at temperature T_1. The gas expands reversibly and isothermally . During this process, the system absorbs energy as heat (Q_1) from the source. The system changes its state from 1 to 2 on the $P-v$ diagram.

$$Q_1 = (U_2 - U_1) + W_{1-2}$$

where, for an ideal gas, $U_2 - U_1$

Reversible Adiabatic Expansion

In the second process, the cylinder head is insulated and the gas is allowed to expand till its temperature is equal to the sink temperature T_2 . The system thus reaches state 3. This is a reversible adiabatic process.

$$0 = (U_3 - U_2) + W_{2-3}$$

Reversible Isothermal Heat Rejection

In the next process, the system is brought into contact with the sink which is at a temperature T_2 . The heat Q_2 leaves the system and the internal energy further decreases

$$-Q_2 = (U_4 - U_3) + W_{3-4}$$

where, only for an ideal gas, $U_4 - U_3$

Reversible Adiabatic Compression

In the next process, the gas is compressed reversibly and adiabatically till it reaches the initial state 1, thus, completing the cycle.

$$0 = U_1 - U_4 - W_{4-1}$$

Summing up all the processes, one can write

$$Q_1 - Q_2 = (W_{1-2} + W_{2-3}) + (W_{3-4} + W_{4-1})$$

or,

$$\sum_{cycle} Qnet = \sum_{cycle} Wnet$$

The thermal efficiency,

$$\eta = \frac{W}{Q_1} = \frac{Q_1 - Q_2}{Q_1} = 1 - \frac{Q_2}{Q_1}$$

Efficiency of Carnot Engine Using Ideal Gas

1-2: A reversible isothermal expansion with heat addition

$$Q_1 = \int_1^2 Pdv = RT_1\, In\left(\frac{v_2}{v_1}\right)$$

2-3: A reversible adiabatic expansion

$$W = c_v (T_1 - T_2)$$

3-4: A reversible isothermal compression with heat rejection

$$Q_2 = \int_3^4 Pdv = RT_2\, In\left(\frac{v_4}{v_3}\right)$$

4-1: A reversible adiabatic compression

$$W = \int_4^1 du = c_v (T_2 - T_1)$$

$$\Sigma \ddot{u}\ddot{u}\ddot{u}\ddot{u}\ddot{u}\ddot{u}\ddot{u}_1 \left(\frac{\hat{v}_2}{\hat{v}_1}\right) + {}_v(\,_1 - \,_2) + \quad _2 \left(\frac{v_4}{v_3}\right) + {}_v(\,_2 - \,_1)$$

$$= RT_1\, In\left(\frac{v_2}{v_1}\right) + RT_2\, In\left(\frac{v_4}{v_3}\right)$$

Energy absorbed as heat

$$Q_1 = RT_1\, In(v_2/v_1)$$

Thermal efficiency,

$$\eta = 1 + \frac{RT_2\ In(v_4/v_3)}{RT_1\ In(v_2/v_1)}$$

Here, for the ideal gases we can write

$$\frac{T_2}{T_1} = \left(\frac{v_2}{v_3}\right)^{\gamma-1}$$

$$\frac{T_2}{T_1} = \left(\frac{v_1}{v_4}\right)^{\gamma-1}$$

Also,

$$\frac{v_2}{v_3} = \frac{v_1}{v_4}$$

or,

$$\frac{v_2}{v_1} = \frac{v_3}{v_4}$$

So,

$$\eta = 1 + \frac{T_2\ In(v_2/v_1)}{T_1\ In(v_2/v_1)} = 1 - \frac{T_2}{T_1}$$

Carnot Cycle

The Carnot cycle is a theoretical thermodynamic cycle proposed by Nicolas Léonard Sadi Carnot in 1824 and expanded upon by others in the 1830s and 1840s. It provides an upper limit on the efficiency that any classical thermodynamic engine can achieve during the conversion of heat into work, or conversely, the efficiency of a refrigeration system in creating a temperature difference (e.g. refrigeration) by the application of work to the system. It is not an actual thermodynamic cycle but is a theoretical construct.

Every single thermodynamic system exists in a particular state. When a system is taken through a series of different states and finally returned to its initial state, a thermodynamic cycle is said to have occurred. In the process of going through this cycle, the system may perform work on its surroundings, thereby acting as a heat engine. A system undergoing a Carnot cycle is called a Carnot heat engine, although such a "perfect" engine is only a theoretical construct and cannot be built in practice. However, a microscopic Carnot heat engine has been designed and run.

Essentially, there are two systems at temperatures T_h and T_c (hot and cold respectively), which are so large that their temperatures are practically unaffected by a single cycle. As such, they are called "heat reservoirs". Since the cycle is reversible, there is no generation of entropy during the cycle; entropy is conserved. During the cycle, an arbitrary amount of entropy ΔS is extracted from the hot reservoir, and deposited in the cold reservoir. Since there is no volume change in either reservoir, they do no work, and during the cycle, an amount of energy $T_h\Delta S$ is extracted from the hot reservoir and a smaller amount of energy $T_c\Delta S$ is deposited in the cold reservoir. The difference in the two energies $(T_h-T_c)\Delta S$ is equal to the work done by the engine.

Stages

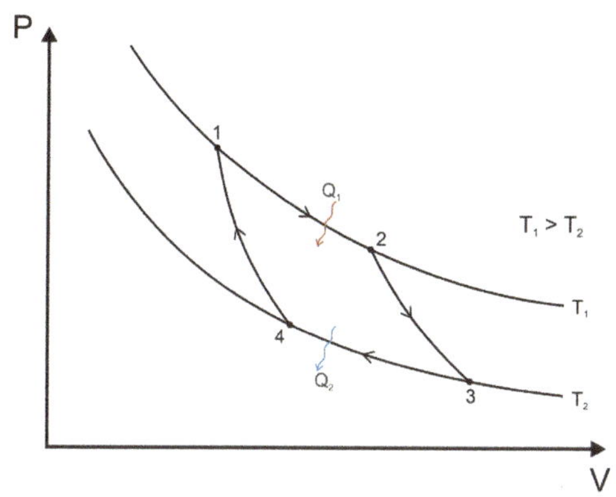

A Carnot cycle illustrated on a PV diagram to illustrate the work done.

The Carnot cycle when acting as a heat engine consists of the following steps:

1. Reversible isothermal expansion of the gas at the "hot" temperature, T_1 (isothermal heat addition or absorption). During this step (1 to 2 on Figure, A to B in Figure) the gas is allowed to expand and it does work on the surroundings. The temperature of the gas does not change during the process, and thus the expansion is isothermal. The gas expansion is propelled by absorption of heat energy Q_1 from the high temperature reservoir and results in an increase of entropy of the gas in the amount $\Delta S_1 = Q_1 / T_1$.

2. Isentropic (reversible adiabatic) expansion of the gas (isentropic work output). For this step (2 to 3 on Figure, B to C in Figure) the mechanisms of the engine are assumed to be thermally insulated, thus they neither gain nor lose heat (an adiabatic process). The gas continues to expand, doing work on the surroundings, and losing an amount of internal energy equal to the work that leaves the system. The gas expansion causes it to cool to the "cold" temperature, T_2. The entropy remains unchanged.

3. Reversible isothermal compression of the gas at the "cold" temperature, T_2. (isothermal heat rejection) (3 to 4 on Figure, C to D on Figure) Now the surroundings do work on the gas, causing an amount of heat energy Q_2 to leave the

system to the low temperature reservoir and the entropy of the system decreases in the amount $\Delta S_2 = Q_2 / T_2$. (This is the same amount of entropy absorbed in step 1, as can be seen from the Clausius inequality.)

4. Isentropic compression of the gas (isentropic work input). (4 to 1 on Figure, D to A on Figure) Once again the mechanisms of the engine are assumed to be thermally insulated, and frictionless, hence reversible. During this step, the surroundings do work on the gas, increasing its internal energy and compressing it, causing the temperature to rise to T_1 due solely to the work added to the system, but the entropy remains unchanged. At this point the gas is in the same state as at the start of step 1.

In this case,

$$\Delta S_1 = \Delta S_2,$$

or,

$$Q_1 / T_1 = Q_2 / T_2.$$

This is true as Q_2 and T_2 are both lower and in fact are in the same ratio as Q_1 / T_1.

The Pressure-volume Graph

When the Carnot cycle is plotted on a pressure volume diagram, the isothermal stages follow the isotherm lines for the working fluid, adiabatic stages move between isotherms and the area bounded by the complete cycle path represents the total work that can be done during one cycle.

Properties and Significance

The Temperature-entropy Diagram

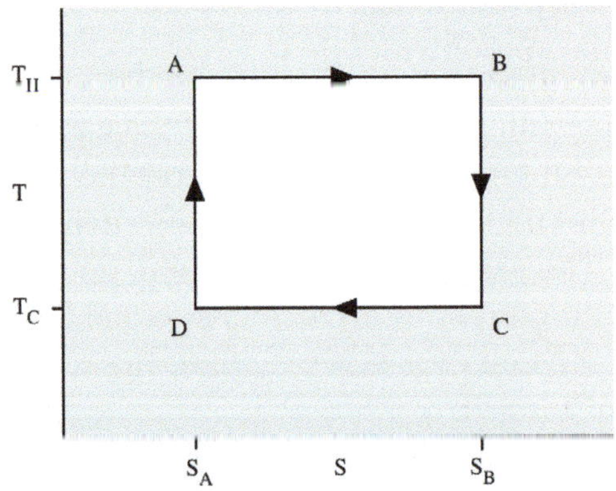

A Carnot cycle acting as a heat engine, illustrated on a temperature-entropy diagram. The cycle takes place between a hot reservoir at temperature T_H and a cold reservoir at temperature T_C. The vertical axis is temperature, the horizontal axis is entropy.

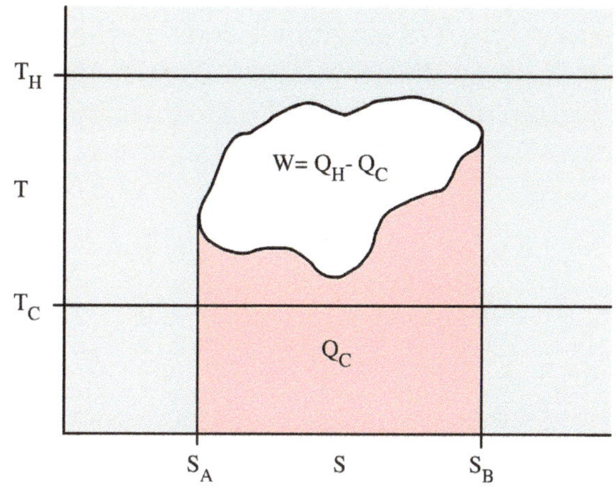

A generalized thermodynamic cycle taking place between a hot reservoir at temperature T_H and a cold reservoir at temperature T_C. By the second law of thermodynamics, the cycle cannot extend outside the temperature band from T_C to T_H. The area in red Q_C is the amount of energy exchanged between the system and the cold reservoir. The area in white W is the amount of work energy exchanged by the system with its surroundings. The amount of heat exchanged with the hot reservoir is the sum of the two. If the system is behaving as an engine, the process moves clockwise around the loop, and moves counter-clockwise if it is behaving as a refrigerator. The efficiency to the cycle is the ratio of the white area (work) divided by the sum of the white and red areas (heat absorbed from the hot reservoir).

The behaviour of a Carnot engine or refrigerator is best understood by using a temperature-entropy diagram (TS diagram), in which the thermodynamic state is specified by a point on a graph with entropy (S) as the horizontal axis and temperature (T) as the vertical axis. For a simple closed system (control mass analysis), any point on the graph will represent a particular state of the system. A thermodynamic process will consist of a curve connecting an initial state (A) and a final state (B). The area under the curve will be:

$$Q = \int_A^B T\,dS$$

which is the amount of thermal energy transferred in the process. If the process moves to greater entropy, the area under the curve will be the amount of heat absorbed by the system in that process. If the process moves towards lesser entropy, it will be the amount of heat removed. For any cyclic process, there will be an upper portion of the cycle and a lower portion. For a clockwise cycle, the area under the upper portion will be the thermal energy absorbed during the cycle, while the area under the lower portion will be the thermal energy removed during the cycle. The area inside the cycle will then be the difference between the two, but since the internal energy of the system must have returned to its initial value, this difference must be the amount of work done by the system over the cycle. Referring to figure, mathematically, for a reversible process we may write the amount of work done over a cyclic process as:

$$W = \oint PdV = \oint(dQ - dU) = \oint(TdS - dU) = \oint TdS - \oint dU = \oint TdS$$

Since dU is an exact differential, its integral over any closed loop is zero and it follows that the area inside the loop on a T-S diagram is equal to the total work performed if the loop is traversed in a clockwise direction, and is equal to the total work done on the system as the loop is traversed in a counterclockwise direction.

The Carnot Cycle

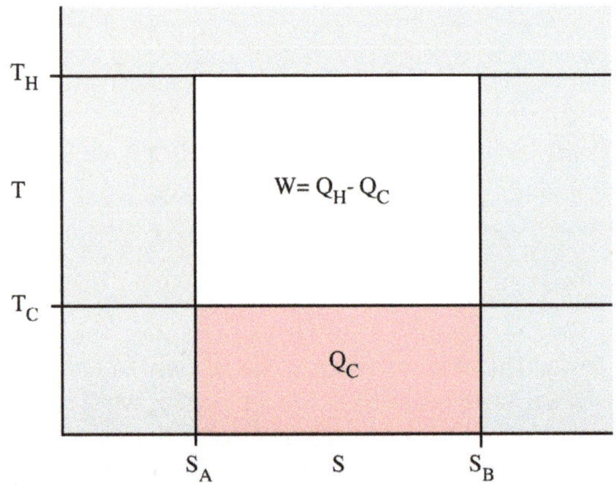

A Carnot cycle taking place between a hot reservoir at temperature T_H and a cold reservoir at temperature T_C.

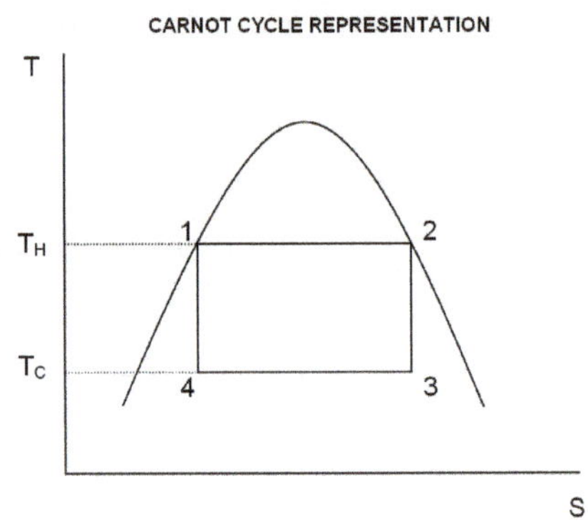

A visualization of the Carnot cycle

Evaluation of the above integral is particularly simple for the Carnot cycle. The amount of energy transferred as work is

$$W = \oint PdV = \oint TdS = (T_H - T_C)(S_B - S_A)$$

The total amount of thermal energy transferred from the hot reservoir to the system will be

$$Q_H = T_H(S_B - S_A)$$

and the total amount of thermal energy transferred from the system to the cold reservoir will be

$$Q_C = T_C(S_B - S_A)$$

The efficiency η is defined to be:

$$\eta = \frac{W}{Q_H} = 1 - \frac{T_C}{T_H}$$

where

W is the work done by the system (energy exiting the system as work),

Q_C is the heat taken from the system (heat energy leaving the system),

Q_H is the heat put into the system (heat energy entering the system),

T_C is the absolute temperature of the cold reservoir, and

T_H is the absolute temperature of the hot reservoir.

S_B is the maximum system entropy

S_A is the minimum system entropy

This definition of efficiency makes sense for a heat engine, since it is the fraction of the heat energy extracted from the hot reservoir and converted to mechanical work. A Rankine cycle is usually the practical approximation.

The Reversed Carnot Cycle

The Carnot heat-engine cycle described is a totally reversible cycle. That is, all the processes that comprise it can be reversed, in which case it becomes the Carnot refrigeration cycle. This time, the cycle remains exactly the same except that the directions of any heat and work interactions are reversed. Heat is absorbed from the low-temperature reservoir, heat is rejected to a high-temperature reservoir, and a work input is required to accomplish all this. The P-V diagram of the reversed Carnot cycle is the same as for the Carnot cycle except that the directions of the processes are reversed.

Carnot's Theorem

It can be seen from the above diagram, that for any cycle operating between temperatures T_H and T_C, none can exceed the efficiency of a Carnot cycle.

Carnot's theorem is a formal statement of this fact: *No engine operating between two heat reservoirs can be more efficient than a Carnot engine operating between those same reservoirs.* Thus, Equation 3 gives the maximum efficiency possible for any engine using the corresponding temperatures. A corollary to Carnot's theorem states that: *All reversible engines operating between the same heat reservoirs are equally efficient.*

Rearranging the right side of the equation gives what may be a more easily understood form of the equation. Namely that the theoretical maximum efficiency of a heat engine equals the difference in temperature between the hot and cold reservoir divided by the absolute temperature of the hot reservoir. To find the absolute temperature in kelvins, add 273.15 degrees to the Celsius temperature. Looking at this formula an interesting fact becomes apparent. Lowering the temperature of the cold reservoir will have more effect on the ceiling efficiency of a heat engine than raising the temperature of the hot reservoir by the same amount. In the real world, this may be difficult to achieve since the cold reservoir is often an existing ambient temperature.

In other words, maximum efficiency is achieved if and only if no new entropy is created in the cycle, which would be the case if e.g. friction leads to dissipation of work into heat. In that case the cycle is not reversible and the Clausius theorem becomes an inequality rather than an equality. Otherwise, since entropy is a state function, the required dumping of heat into the environment to dispose of excess entropy leads to a (minimal) reduction in efficiency. So earlier equation gives the efficiency of any reversible heat engine.

Corollary to Carnot's theorem states that: *All reversible engines operating between the same heat reservoirs are equally efficient.* Rearranging the right side of the equation gives what may be a more easily understood form of the equation. Namely that the theoretical maximum efficiency of a heat engine equals the difference in temperature between the hot and cold reservoir divided by the absolute temperature of the hot reservoir. To find the absolute temperature in kelvin, add 273.15 degrees to the Celsius temperature. Looking at this formula an interesting fact becomes apparent. Lowering the temperature of the cold reservoir will have more effect on the ceiling efficiency of a heat engine than raising the temperature of the hot reservoir by the same amount. In the real world, this may be difficult to achieve since the cold reservoir is often an existing ambient temperature. In other words, maximum efficiency is achieved if and only if no new entropy is created in the cycle. Otherwise, since entropy is a state function, the required dumping of heat into the environment to dispose of excess entropy leads to a reduction in efficiency. This relation transforms Carnot's inequality into an exact equality that applies to an arbitrary heat engine coupled to two heat reservoirs and operating at an arbitrary rate.

In mesoscopic heat engines, work per cycle of operation fluctuates due to thermal noise. For the case when work and heat fluctuations are counted, there is exact equality that relates average of exponents of work performed by any heat engine and the heat transfer from the hotter heat bath.

Efficiency of Real Heat Engines

Carnot realized that in reality it is not possible to build a thermodynamically reversible engine, so real heat engines are even less efficient than indicated by Equation 3. In addition, real engines that operate along this cycle are rare. Nevertheless, Equation 3 is extremely useful for determining the maximum efficiency that could ever be expected for a given set of thermal reservoirs.

Although Carnot's cycle is an idealisation, the expression of Carnot efficiency is still useful. Consider the average temperatures,

$$\langle T_H \rangle = \frac{1}{\Delta S} \int_{Q_{in}} T dS$$

$$\langle T_C \rangle = \frac{1}{\Delta S} \int_{Q_{out}} T dS$$

at which heat is input and output, respectively. Replace T_H and T_C in Equation (3) by $\langle T_H \rangle$ and $\langle T_C \rangle$ respectively.

For the Carnot cycle, or its equivalent, the average value $\langle T_H \rangle$ will equal the highest temperature available, namely T_H, and $\langle T_C \rangle$ the lowest, namely T_C. For other less efficient cycles, $\langle T_H \rangle$ will be lower than T_H, and $\langle T_C \rangle$ will be higher than T_C. This can help illustrate, for example, why a reheater or a regenerator can improve the thermal efficiency of steam power plants—and why the thermal efficiency of combined-cycle power plants (which incorporate gas turbines operating at even higher temperatures) exceeds that of conventional steam plants. The first prototype of the diesel engine was based on the Carnot cycle.

Carnot Heat Engine

A Carnot heat engine is an engine that operates on the reversible Carnot cycle. The basic model for this engine was developed by Nicolas Léonard Sadi Carnot in 1824. The Carnot engine model was graphically expanded upon by Benoît Paul Émile Clapeyron in 1834 and mathematically elaborated upon by Rudolf Clausius in 1857 from which the concept of entropy emerged.

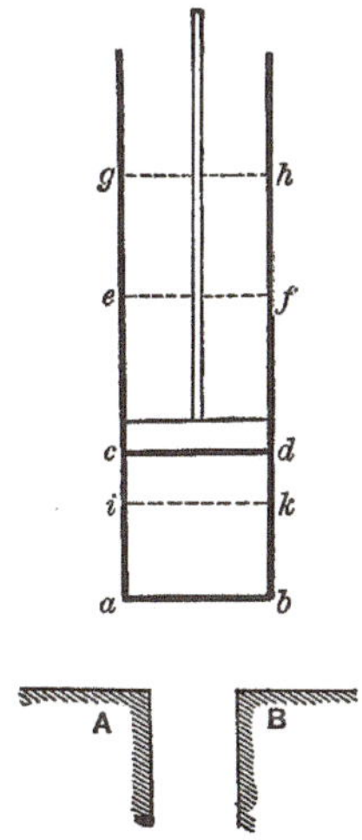

Axial cross section of Carnot's heat engine. In this diagram, *abcd* is a cylindrical vessel, *cd* is a movable piston, and *A* and *B* are constant–temperature bodies. The vessel may be placed in contact with either body or removed from both (as it is here).

Every thermodynamic system exists in a particular state. A thermodynamic cycle occurs when a system is taken through a series of different states, and finally returned to its initial state. In the process of going through this cycle, the system may perform work on its surroundings, thereby acting as a heat engine.

A heat engine acts by transferring energy from a warm region to a cool region of space and, in the process, converting some of that energy to mechanical work. The cycle may also be reversed. The system may be worked upon by an external force, and in the process, it can transfer thermal energy from a cooler system to a warmer one, thereby acting as a refrigerator or heat pump rather than a heat engine.

Carnot's Diagram

In the adjacent diagram, from Carnot's 1824 work, *Reflections on the Motive Power of Fire*, there are "two bodies *A* and *B*, kept each at a constant temperature, that of *A* being higher than that of *B*. These two bodies to which we can give, or from which we can remove the heat without causing their temperatures to vary, exercise the functions of two unlimited reservoirs of caloric. We will call the first the furnace and the second the refrigerator." Carnot then explains how we can obtain motive power, i.e., "work", by carrying a certain quantity of heat from body *A* to body *B*. It also acts as a cooler and hence can also act as a Refrigerator.

Modern Diagram

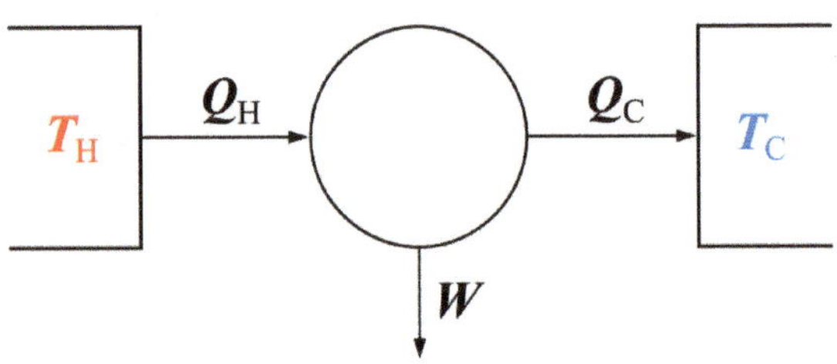

Carnot engine diagram (modern) - where an amount of heat Q_H flows from a high temperature T_H furnace through the fluid of the "working body" (working substance) and the remaining heat Q_C flows into the cold sink T_C, thus forcing the working substance to do mechanical work W on the surroundings, via cycles of contractions and expansions.

The previous image shows the original piston-and-cylinder diagram used by Carnot in discussing his ideal engines. The figure at right shows a block diagram of a generic heat engine, such as the Carnot engine. In the diagram, the "working body" (system), a term introduced by Clausius in 1850, can be any fluid or vapor body through which heat Q can be introduced or transmitted to produce work. Carnot had postulated that the fluid body could be any substance capable of expansion, such as vapor of water, vapor of alcohol, vapor of mercury, a permanent gas, or air, etc. Although, in these early years, engines came in a number of configurations, typically Q_H was supplied by a boiler, wherein water was boiled over a furnace; Q_C was typically supplied by a stream of cold flowing water in the form of a condenser located on a separate part of the engine. The output work *W* here is the movement of the piston as it is used to turn a crank-arm, which was then typically used to turn a pulley so to lift water out of flooded salt mines. Carnot defined work as “weight lifted through a height”.

Carnot Cycle

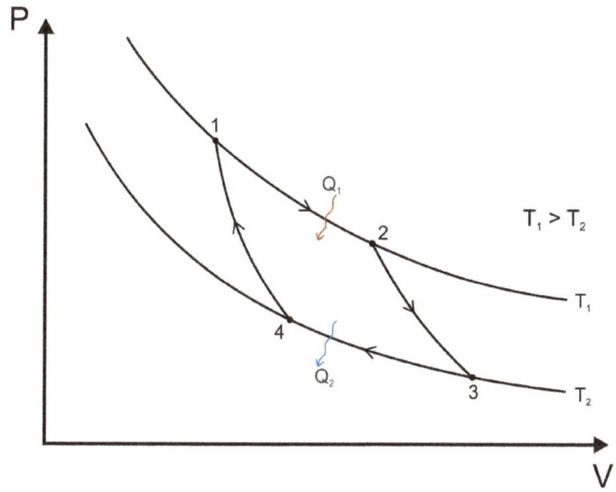

A Carnot cycle illustrated on a PV diagram to illustrate the work done.

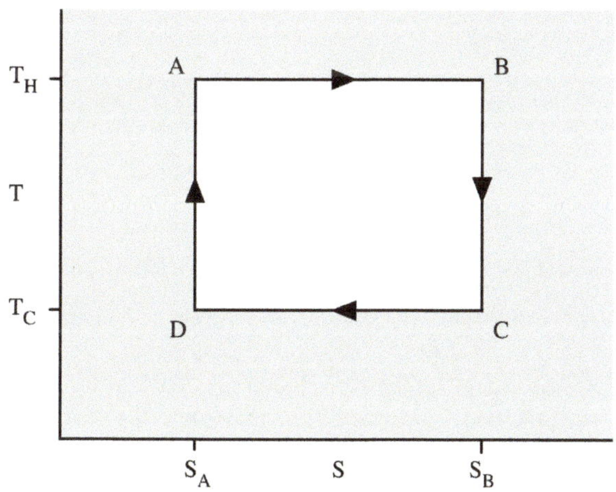

A Carnot cycle acting as a heat engine, illustrated on a temperature-entropy diagram. The cycle takes place between a hot reservoir at temperature T_H and a cold reservoir at temperature T_C. The vertical axis is temperature, the horizontal axis is entropy.

The Carnot cycle when acting as a heat engine consists of the following steps:

1. Reversible isothermal expansion of the gas at the “hot” temperature, T_H (isothermal heat addition or absorption). During this step (1 to 2 on Figure, *A* to *B* in Figure) the gas is allowed to expand and it does work on the surroundings. The temperature of the gas does not change during the process, and thus the

expansion is isothermic. The gas expansion is propelled by absorption of heat energy Q_1 and of entropy $\Delta S_H = Q_H / T_H$ from the high temperature reservoir.

2. Isentropic (reversible adiabatic) expansion of the gas (isentropic work output). For this step (2 to 3 on Figure, *B* to *C* in Figure) the piston and cylinder are assumed to be thermally insulated, thus they neither gain nor lose heat. The gas continues to expand, doing work on the surroundings, and losing an equivalent amount of internal energy. The gas expansion causes it to cool to the "cold" temperature, T_C. The entropy remains unchanged.

3. Reversible isothermal compression of the gas at the "cold" temperature, T_C. (isothermal heat rejection) (3 to 4 on Figure, *C* to *D* on Figure) Now the surroundings do work on the gas, causing an amount of heat energy Q_2 and of entropy $\Delta S_C = Q_C / T_C$ to flow out of the gas to the low temperature reservoir. (This is the same amount of entropy absorbed in step 1.)

4. Isentropic compression of the gas (isentropic work input). (4 to 1 on Figure, *D* to *A* on Figure) Once again the piston and cylinder are assumed to be thermally insulated. During this step, the surroundings do work on the gas, increasing its internal energy and compressing it, causing the temperature to rise to T_H. The entropy remains unchanged. At this point the gas is in the same state as at the start of step 1.

Carnot's Theorem

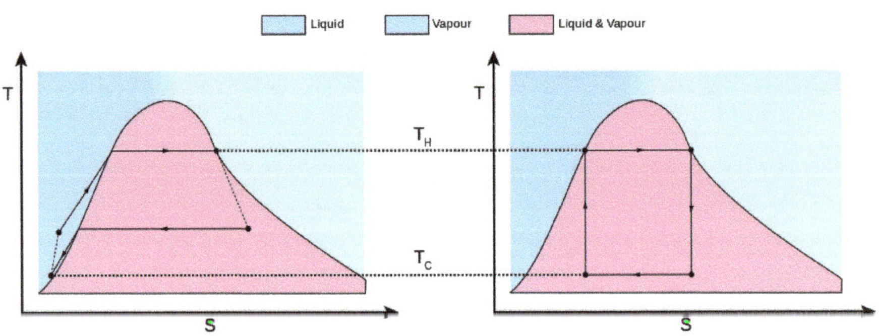

Real ideal engines (left) compared to the Carnot cycle (right). The entropy of a real material changes with temperature. This change is indicated by the curve on a T-S diagram. For this figure, the curve indicates a vapor-liquid equilibrium. Irreversible systems and losses of heat (for example, |due to friction) prevent the ideal from taking place at every step.

Carnot's theorem is a formal statement of this fact: *No engine operating between two heat reservoirs can be more efficient than a Carnot engine operating between the same reservoirs.*

$$= \frac{W}{Q_H} = 1 - \frac{T}{T_H}$$

Explanation

This maximum efficiency η_I is defined as above:

W is the work done by the system (energy exiting the system as work),

Q_H is the heat put into the system (heat energy entering the system),

T_C is the absolute temperature of the cold reservoir, and

T_H is the absolute temperature of the hot reservoir.

A corollary to Carnot's theorem states that: All reversible engines operating between the same heat reservoirs are equally efficient.

It is easily shown that the efficiency η is maximum when the entire cyclic process is a reversible process. This means the total entropy of the total system consisting of the three parts: the entropy of the hot furnace, the entropy of the "working fluid" of the Heat engine, and the entropy of the cold sink. The total entropy of the system remains constant when the "working fluid" completes one cycle and returns to its original state. (In the general case, the total entropy of this combined system would increase in a general irreversible process).

Since the "working fluid" comes back to the same state after one cycle, and entropy of the system is a state function; the change in entropy of the "working fluid" system is 0. Thus, it implies that the total entropy change of the furnace and sink is zero, for the process to be reversible and the efficiency of the engine to be maximum.

The Coefficient of Performance (COP) of the heat engine is the reciprocal of its efficiency.

Efficiency of Real Heat Engines

For a real heat engine, the total thermodynamic process is generally irreversible. The working fluid is brought back to its initial state after one cycle, and thus the change of entropy of the fluid system is 0, but the sum of the entropy changes in the hot and cold reservoir in this one cyclical process is greater than 0.

The internal energy of the fluid is also a state variable, so its total change in one cycle is 0. So the total work done by the system W, is equal to the heat put into the system Q_H minus the heat taken out Q_C.

$$W = Q_H - Q_C$$

For real engines, sections 1 and 3 of the Carnot Cycle; in which heat is absorbed by the "working fluid" from the hot reservoir, and released by it to the cold reservoir, respectively; no longer remain ideally reversible, and there is a temperature differential between the temperature of the reservoir and the temperature of the fluid while heat exchange takes place.

During heat transfer from the hot reservoir at T_H to the fluid, the fluid would have a slightly lower temperature than T_H, and the process for the fluid may not necessarily remain isothermal. Let ΔS_H be the total entropy change of the fluid in the process of intake of heat.

$$\Delta S_H = \int_{Q_{in}} \frac{dQ_H}{T}$$

where the temperature of the fluid T is always slightly lesser than T_H, in this process.

So, one would get

$$\frac{Q_H}{T_H} = \frac{\int dQ_H}{T_H} \leq \Delta S_H$$

Similarly, at the time of heat injection from the fluid to the cold reservoir one would have, for the magnitude of total entropy change ΔS_C of the fluid in the process of expelling heat:

$$\Delta S_C = \int_{Q_{out}} \frac{dQ_C}{T} \leq \frac{\int dQ_C}{T_C} = \frac{Q_C}{T_C}$$

where, during this process of transfer of heat to the cold reservoir, the temperature of the fluid T is always slightly greater than T_C.

We have only considered the magnitude of the entropy change here. Since the total change of entropy of the fluid system for the cyclic process is 0, we must have

$$\Delta S_H = \Delta S_C$$

The previous three equations combine to give:

$$\frac{Q_C}{T_C} \geq \frac{Q_H}{T_H}$$

Equations $W = Q_H - Q_C$ and $\frac{Q_C}{T_C} \geq \frac{Q_H}{T_H}$ combine to give

$$\frac{W}{Q_H} \leq (1 - \frac{T_C}{T_H})$$

Hence,

$$\eta \leq \eta_I$$

where $\eta = \frac{W}{Q_H}$ is the efficiency of the real engine, and η_I is the efficiency of the Carnot engine working between the same two reservoirs at the temperatures T_H and T_C. For the Carnot engine, the entire process is 'reversible', and earlier equation is an equality.

Hence, the efficiency of the real engine is always less than the ideal Carnot engine.

Earlier equation signifies that the total entropy of the total system(the two reservoirs + fluid) increases for the real engine, because the entropy gain of the cold reservoir as Q_C flows into it at the fixed temperature T_C, is greater than the entropy loss of the hot reservoir as Q_H leaves it at its fixed temperature T_H.. The inequality in earlier equation is essentially the statement of the Clausius theorem.

According to the second theorem, "The efficiency of the Carnot engine is independent of the nature of the working substance".

Substitution of a Reversible Process by the Reversible Isothermal and Reversible Adiabatic Processes:

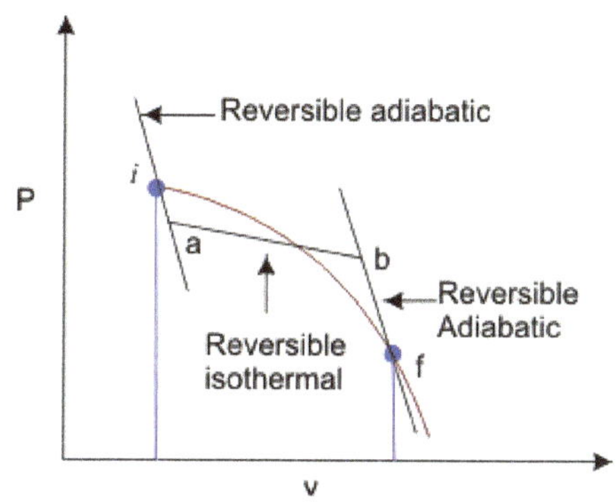

Let a system be taken from an equilibrium state i to another equilibrium state f by following the reversible path *i* - *f*. Refer to figure Let a reversible adiabatic *i* - *a* and another reversible *b* - *f* be drawn in such a manner that a and b can be joined by a reversible isotherm.

Also, these reversible adiabatic and reversible isothermal lines are such that the area under $iabf$ is equal to the area under $i = f$.

Applying the first law for $i = f$,

$$Q_i - f = U_f - U_i + W_{if}$$

Applying the first law for $iabf$,

$$Q_{iabf} = U_f - U_i + W_{iabf}$$

We know $W_{if} = W_{iabf}$ (area under the curves are same)

From above given two equations,

$$Q_i - f = Q_{iabf} = Q_{ia} + Q_{ab} + Q_{bf}$$

Since Q_{ia} and Q_{bf} are zero

$$Q_{i-f} = Q_{ab}$$

Thus, any reversible path may be substituted by a reversible adiabatic, a reversible isotherm and a reversible adiabatic between the same end states such that the heat transferred during the isothermal process is the same as that transferred during the original process.

Clausius Inequality

Consider a system undergoing a reversible cycle. The given cycle may be sub-divided by drawing a family of reversible adiabatic lines. Every two adjacent adiabatic lines may be joined by two reversible isotherms.

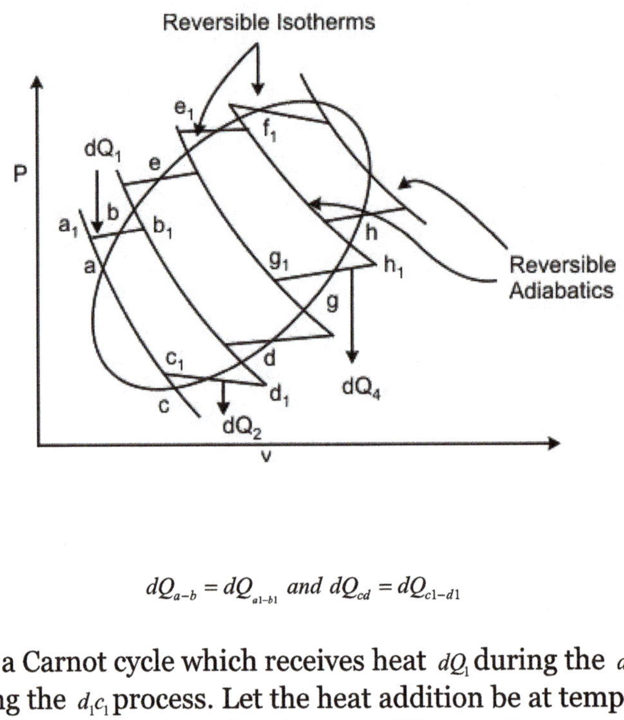

Now,

$$dQ_{a-b} = dQ_{a1-b1} \text{ and } dQ_{cd} = dQ_{c1-d1}$$

Also, $a_1 - b_1 - d_1 - c_1$ is a Carnot cycle which receives heat dQ_1 during the a_1b_1 process and rejects heat dQ_2 during the d_1c_1 process. Let the heat addition be at temperature T_1 and the heat rejection be at temperature T_2. Then it is possible to write,

$$\eta = 1 - \frac{dQ_2}{dQ_1} = 1 - \frac{T_2}{T_1}$$

and

$$\frac{dQ_1}{dQ_2} = \frac{T_2}{T_1}$$

or,

$$\frac{dQ_1}{T_1} = \frac{dQ_2}{T_2}$$

Since dQ_2 is negative, it reduces to

$$\frac{dQ_1}{T_1} = \frac{dQ_2}{T_2} = 0$$

Similarly for the cycle $e_1 f_1 h_1 g_1$

$$\frac{dQ_3}{T_3} = \frac{dQ_4}{T_4} = 0$$

If similar equations are written for all the elementary cycles, then

$$\frac{dQ_1}{T_1} + \frac{dQ_2}{T_2} + \frac{dQ_3}{T_3} + \frac{dQ_4}{T_4} = 0$$

or,

$$\oint_R \frac{dQ}{T} = 0$$

Let us go back to the cycle $e_1 f_1 h_1 g_1$

$$\eta_R = 1 - \frac{dQ_2}{dQ_1} = 1 - \frac{T_2}{T_1}$$

Now $\eta_{IR} < \eta_R$ where, $\eta_{IR} = \left[1 - \frac{dQ_2}{dQ_1}\right]$ and this is not equal to $\left[1 - \frac{T_2}{T_1}\right]$.

For the irreversible cycle,

$$1 - \frac{dQ_2}{dQ_1} < 1 - \frac{T_2}{T_1}$$

or,

$$\frac{dQ_2}{T_2} > \frac{dQ_1}{T_1} \; and \; \frac{dQ_1}{T_1} + \frac{dQ_2}{T_2} < 0$$

because dQ_2 is negative.

Similarly, for the irreversible cycle $e_1 f_1 h_1 g_1$

$$\frac{dQ_3}{T_3} + \frac{dQ_4}{T_4} < 0$$

Summing up all elementary cycles

$$\oint_{IR} \frac{dQ}{T} < 0$$

The above two conclusions about reversible and irreversible cycles can be generalized as

$$\oint \frac{dQ}{T} \leq 0$$

The equality holds good for a reversible cycle and the inequality holds good for an irreversible cycles. The complete expression is known as Clausius Inequality.

References

- N. A. Sinitsyn (2011). "Fluctuation Relation for Heat Engines". J. Phys. A: Math. Theor. 44: 405001. Bibcode:2011JPhA...44N5001S. arXiv:1111.7014. doi:10.1088/1751-8113/44/40/405001
- Kittel, C. Kroemer, H. (1980). Thermal Physics, (first edition by Kittel alone 1969), second edition, W. H. Freeman, San Francisco, ISBN 0-7167-1088-9, pp. 49, 227
- Lebowitz, Joel L. (September 1993). "Boltzmann's Entropy and Time's Arrow" (PDF). Physics Today. 46 (9): 32–38. Bibcode:1993PhT....46i..32L. doi:10.1063/1.881363. Retrieved 2013-02-22
- Gyenis, Balazs (2017). "Maxwell and the normal distribution: A colored story of probability, independence, and tendency towards equilibrium". Studies in History and Philosophy of Modern Physics. 57: 53–65. doi:10.1016/j.shpsb.2017.01.001
- Sychev, V. V. (1991). The Differential Equations of Thermodynamics. Taylor & Francis. ISBN 978-1-56032-121-7. Retrieved 2012-11-26
- Hawking, SW (1985). "Arrow of time in cosmology". Phys. Rev. D. 32 (10): 2489–2495. Bibcode:1985PhRvD..32.2489H. doi:10.1103/PhysRevD.32.2489. Retrieved 2013-02-15
- Tro, N. J. (2008). Chemistry. A Molecular Approach, Pearson/Prentice Hall, Upper Saddle River NJ, ISBN 0-13-100065-9, p. 246
- Grubbström, Robert W. (2007). "An Attempt to Introduce Dynamics Into Generalised Exergy Considerations". Applied Energy. 84: 701–718. doi:10.1016/j.apenergy.2007.01.003
- Atkins, P., de Paula, J. (1978/2010). Physical Chemistry, (first edition 1978), ninth edition 2010, Oxford University Press, Oxford UK, ISBN 978-0-19-954337-3, p. 54
- Gislason, E. A.; Craig, N. C. (2005). "Cementing the foundations of thermodynamics:comparison of system-based and surroundings-based definitions of work and heat". J. Chem. Thermodynamics. 37: 954–966. doi:10.1016/j.jct.2004.12.012
- Rosenberg, R. M. (2010). "From Joule to Caratheodory and Born: A conceptual evolution of the first law of thermodynamics". J. Chem. Edu. 87: 691–693. Bibcode:2010JChEd..87..691R. doi:10.1021/ed1001976
- J. S. Dugdale (1996). Entropy and its Physical Meaning. Taylor & Francis. p. 13. ISBN 0-7484-0569-0. This law is the basis of temperature
- Cropper, W. H. (1986). "Rudolf Clausius and the road to entropy". Am. J. Phys. 54: 1068–1074. Bibcode:1986AmJPh..54.1068C. doi:10.1119/1.14740

- Truesdell, C., Muncaster, R.G. (1980). Fundamentals of Maxwell's Kinetic Theory of a Simple Monatomic Gas, Treated as a Branch of Rational Mechanics, Academic Press, New York, ISBN 0-12-701350-4, p. 15
- Hess, H. (1840). "Thermochemische Untersuchungen". Annalen der Physik und Chemie. 126 (6): 385–404. Bibcode:1840AnP...126..385H. doi:10.1002/andp.18401260620

4

Various Thermodynamical Cycles

Energy is the capacity to do work. Energy, on the macroscopic level, can be divided into two kinds, kinetic and potential. The bifurcation of energy on the microscopic level includes translational energy, rotational energy, electronic energy, nuclear energy, etc. The chapter serves as a source to understand the major categories related to energy conversion processes.

Thermal Energy

Thermal radiation in visible light can be seen on this hot metalwork. Thermal energy would ideally be the amount of heat required to warm the metal to its temperature, but this quantity is not well-defined, as there are many ways to obtain a given body at a given temperature, and each of them may require a different amount of total heat input. Thermal energy, unlike internal energy, is therefore not a state function.

In thermodynamics, thermal energy refers to the internal energy present in a system due to its temperature. The concept is not well-defined or broadly accepted in physics or thermodynamics, because the internal energy can be changed without changing the temperature, and there is no way to distinguish which part of a system's internal energy is "thermal". *Thermal energy* is sometimes used loosely as a synonym for more rigorous thermodynamic quantities such as the (entire) internal energy of a system; or for heat or sensible heat which are defined as types of *transfer* of energy (just as work is another type of transfer of energy). Heat and work depend on the way in which an energy transfer occurred, whereas internal energy is a property of the state of a system and can thus be understood even without knowing how the energy got there.

Relation to Heat and Internal Energy

Heat is energy transferred spontaneously from a hotter to a colder system or body. Heat is energy in transfer, not a property of the system; it is not 'contained' within the boundary of the system. On the other hand, internal energy is a property of a system. In an ideal gas, the internal energy is the statistical mean of the kinetic energy of the gas particles, and it is this kinetic motion that is the source and the effect of the transfer of heat across a system boundary. The internal energy of an ideal gas can in this sense be regarded as "thermal energy". In this case, however, thermal energy and internal energy are identical.

Systems that are more complex than ideal gases (such as real gases) can undergo phase transitions. Phase transitions can change the internal energy of the system without changing its temperature. Therefore, the thermal energy cannot be defined solely by the temperature. Thermal energy also cannot be defined by the difference between the internal energy and the net heat transfer in or out of the system, because it is easy to construct thermodynamic cycles where the system begins and ends in exactly the same state, but there is a net flow of heat in or out during the cycle.

For these reasons, the concept of the thermal energy of a system is ill-defined and is not used in thermodynamics.

Historical Context

In an 1847 lecture entitled *On Matter, Living Force, and Heat*, James Prescott Joule characterized various terms that are closely related to thermal energy and heat. He identified the terms latent heat and sensible heat as forms of heat each effecting distinct physical phenomena, namely the potential and kinetic energy of particles, respectively. He described latent energy as the energy of interaction in a given configuration of particles, i.e. a form of potential energy, and the sensible heat as an energy affecting temperature measured by the thermometer due to the thermal energy, which he called the *living force*.

Energy

Energy is often defined as the capacity to produce work. However, this "capacity" has a special significance. The capacity represents a combination of an effort and the change brought about by the effort. However, the effort is exerted in overcoming resistance to a particular type of change.

The effort involved is measured quantitatively as a "driving force" in thermodynamics. A driving force is a property which causes and also controls the direction of change in another property. The quantitative value of this change is called a "displacement". The product of a driving force and its associated displacement represents a quantity of energy, but in thermodynamics this quantity has meaning only in relation to a specifically defined system.

Relative to a particular system there are generally two ways of locating a driving force and the displacement it produces. In one way, the driving force and the displacement are properties of the system and are located entirely within it. The energy calculated from their product represents a change in the internal energy of the system. Similarly, both the driving force and its displacement could be located entirely within the surroundings so that the calculated energy is then a change in the total energy of the surroundings.

In another way, the displacement occurs within the system but the driving force is a property of the surroundings and is applied externally at the system boundary. By definition, the boundary of a system is a region of zero thickness containing no matter at all so that the energy calculated in this way is not a property of matter either in the system or in its surroundings but represents a quantity of energy in transition between the two. In any quantitative application of thermodynamics it is always important to make a careful distinction between energy changes within a system or within its surroundings and energy in transition between them.

Macroscopic Modes of Energy

- Kinetic Energy (KE)

If a body is accelerated from its initial velocity V_1 to final velocity V_2 , the total work done on the body is

$$W = \int_1^2 mVdV = \frac{m\left(V_2^2 - V_1^2\right)}{2} = \frac{\Delta\left((mV^2\right)}{2}$$

The work done on a body in accelerating it from its initial velocity V_1 to a final velocity V_2 , is equal to the change in the kinetic energy of the body. If the body is decelerated from a velocity V_2 to a velocity V_1 by the application of resisting force, the work done by the body is equal to decrease in its kinetic energy.

- Potential Energy (PE)
  - A body of mass m is moved from an initial elevant Z_1 to a final elevation Z_2(Figure)

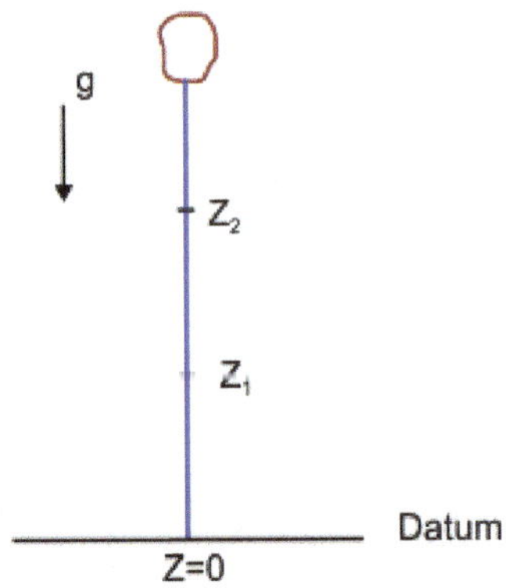

- The force on the body, F = mg
- This force has moved a distance (Z_2 - Z_1). Therefore, the work done on the body

$$W = F(Z_2 - Z_1) = mg(Z_2 - Z_1) = \Delta(mgZ)$$

The kinetic energy and potential energy are also called organized form of energy that can be readily converted into work.

Microscopic Modes of Energy

The microscopic modes of energy refer to the energy stored in the molecular and atomic structure of the system.

The molecules are in random motion. A molecule possesses energy in several forms.

- Translational energy, Rotational energy, Vibrational energy.
- Electronic energy, Chemical energy, Nuclear energy.

If ε represents energy of one molecule, then

$$\varepsilon = \varepsilon_{trans} + \varepsilon_{rot} + \varepsilon_{vib} + \varepsilon_{elec} + \varepsilon_{chem} + \varepsilon_{nucl}$$

If N is the total number of molecules in the system, then the total amount of microscopic form of energy

$$U = N\varepsilon$$

We may also call this as INTERNAL ENERGY of the system. The internal energy, U is due to random motions or disorganized motions of molecules. The internal energy cannot be readily converted into work. One of the major tasks involved in thermodynamics is devising means for converting disorganized internal energy into useful or organized work.

Thermodynamic Equilibrium

Steady State

Under the steady state condition, the properties of the system at any location are independent of time.

Equilibrium

At the state of equilibrium, the properties of the system are uniform and only one value can be assigned to it.

In thermodynamics, equilibrium refers to a state of equilibrium with respect to all possible changes, thermal, mechanical and chemical.

- Thermal equilibrium

A state of thermal equilibrium can be described as one in which the temperature of the system is uniform.

- Mechanical equilibrium

Mechanical equilibrium means there is no unbalanced force. In other words, there is no pressure gradient within the system.

- Chemical equilibrium

The criterian for chemical equilibrium is the equality of chemical potential

$$\mu_i^A = \mu_i^B$$

Superscripts A and B refers to systems and subscript i refers to component

If Gibbs function is given by G, G = U + PV – TS

$$\mu_i = \left(\frac{\partial G}{\partial n_i}\right)_{n_J(J \neq i)}$$

where ni is the number of moles of substance i . The composition of a system does not undergo any change because of chemical reaction.

Thermodynamic Process

Classical thermodynamics considers three main kinds of thermodynamic process: change in a system, cycles in a system, and flow processes.

Defined by change in a system, a thermodynamic process is a passage of a thermodynamic system from an initial to a final state of thermodynamic equilibrium. The initial and final states are the defining elements of the process. The actual course of the process is not the primary concern, and often is ignored. This is the customary default meaning of the term 'thermodynamic process'. In general, during the actual course of a thermodynamic process, the system passes through physical states which are not describable as thermodynamic states, because they are far from internal thermodynamic equilibrium. Such processes are useful for thermodynamic theory.

Defined by a cycle of transfers into and out of a system, a cyclic process is described by the quantities transferred in the several stages of the cycle, which recur unchangingly.

The descriptions of the staged states of the system are not the primary concern. Cyclic processes were important conceptual devices in the early days of thermodynamical investigation, while the concept of the thermodynamic state variable was being developed.

Defined by flows through a system, a flow process is a steady state of flows into and out of a vessel with definite wall properties. The internal state of the vessel contents is not the primary concern. The quantities of primary concern describe the states of the inflow and the outflow materials, and, on the side, the transfers of heat, work, and kinetic and potential energies for the vessel. Flow processes are of interest in engineering.

Kinds of Process

Thermodynamic Process

Defined by change in a system, a thermodynamic process is a passage of a thermodynamic system from an initial to a final state of thermodynamic equilibrium. The initial and final states are the defining elements of the process. The actual course of the process is not the primary concern, and often is ignored. A state of thermodynamic equilibrium endures unchangingly unless it is interrupted by a thermodynamic operation that initiates a thermodynamic process. The equilibrium states are each respectively fully specified by a suitable set of thermodynamic state variables, that depend only on the current state of the system, not the path taken by the processes that produce that state. In general, during the actual course of a thermodynamic process, the system passes through physical states which are not describable as thermodynamic states, because they are far from internal thermodynamic equilibrium. Such a process may therefore be admitted for equilibrium thermodynamics, but not be admitted for non-equilibrium thermodynamics, which primarily aims to describe the continuous passage along the path, at definite rates of progress.

Though not so in general, it is, however, possible, that a process may take place slowly or smoothly enough to allow its description to be usefully approximated by a continuous path of equilibrium thermodynamic states. Then it may be approximately described by a process function that does depend on the path. Such a process may be idealized as a "quasi-static" process, which is infinitely slow, and which is really a theoretical exercise in differential geometry, as opposed to an actually possible physical process; in this idealized case, the calculation may be exact, though the process does not actually occur in nature. Such idealized processes are useful in the theory of thermodynamics.

Cyclic Process

Defined by a cycle of transfers into and out of a system, a cyclic process is described by the quantities transferred in the several stages of the cycle. The descriptions of the staged states of the system may be of little or even no interest. A cycle is a sequence of a small number of thermodynamic processes that indefinitely often repeatedly returns the system to its original state. For this, the staged states themselves are not necessarily described, because it is the transfers that are of interest. It is reasoned that if the cycle can be repeated indefinitely often, then it can be assumed that the states are recurrently unchanged. The condition of the system during the several staged processes may be of even less interest than is the precise nature of the recurrent states. If, however, the several staged processes are idealized and quasi-static, then the cycle is described by a path through a continuous progression of equilibrium states.

Flow Process

Defined by flows through a system, a flow process is a steady state of flow into and out of a vessel with definite wall properties. The internal state of the vessel contents is not the primary concern. The quantities of primary concern describe the states of the inflow and the outflow materials, and, on the side, the transfers of heat, work, and kinetic and potential energies for the vessel. The states of the inflow and outflow materials consist of their internal states, and of their kinetic and potential energies as whole bodies. Very often, the quantities that describe the internal states of the input and output materials are estimated on the assumption that they are bodies in their own states of internal thermodynamic equilibrium. Because rapid reactions are permitted, the thermodynamic treatment may be approximate, not exact.

A Cycle of Quasi-static Processes

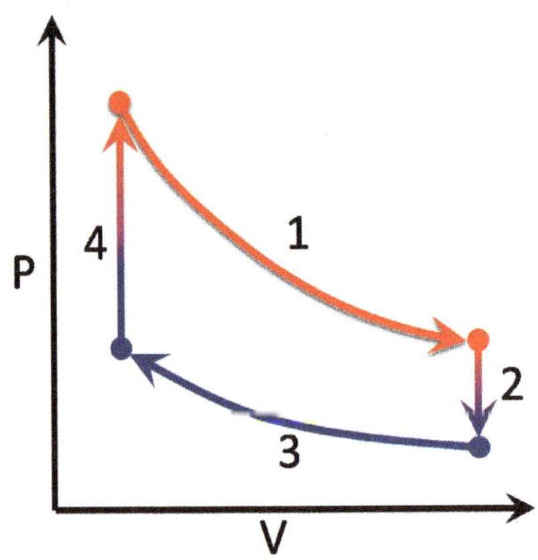

An example of a cycle of idealized thermodynamic processes which make up the Stirling cycle

A quasi-static thermodynamic process can be visualized by graphically plotting the path of idealized changes to the system's state variables. In the example, a cycle consisting of four quasi-static processes is shown. Each process has a well-defined start and end point in the pressure-volume state space. In this particular example, processes 1 and 3 are isothermal, whereas processes 2 and 4 are isochoric. The PV diagram is a particularly useful visualization of a quasi-static process, because the area under the curve of a process is the amount of work done by the system during that process. Thus work is considered to be a process variable, as its exact value depends on the particular path taken between the start and end points of the process. Similarly, heat may be transferred during a process, and it too is a process variable.

Conjugate Variable Processes

It is often useful to group processes into pairs, in which each variable held constant is one member of a conjugate pair.

Pressure - volume

The pressure-volume conjugate pair is concerned with the transfer of mechanical or dynamic energy as the result of work.

- An isobaric process occurs at constant pressure. An example would be to have a movable piston in a cylinder, so that the pressure inside the cylinder is always at atmospheric pressure, although it is separated from the atmosphere. In other words, the system is dynamically connected, by a movable boundary, to a constant-pressure reservoir.
- An isochoric process is one in which the volume is held constant, with the result that the mechanical PV work done by the system will be zero. On the other hand, work can be done isochorically on the system, for example by a shaft that drives a rotary paddle located inside the system. It follows that, for the simple system of one deformation variable, any heat energy transferred to the system externally will be absorbed as internal energy. An isochoric process is also known as an isometric process or an isovolumetric process. An example would be to place a closed tin can of material into a fire. To a first approximation, the can will not expand, and the only change will be that the contents gain internal energy, evidenced by increase in temperature and pressure. Mathematically, $\delta Q = dU$. The system is dynamically insulated, by a rigid boundary, from the environment.

Temperature - entropy

The temperature-entropy conjugate pair is concerned with the transfer of energy, especially for a closed system.

- An isothermal process occurs at a constant temperature. An example would be a closed system immersed in and thermally connected with a large constant-temperature bath. Energy gained by the system, through work done on it, is lost to the bath, so that its temperature remains constant.
- An adiabatic process is a process in which there is no matter or heat transfer,

because a thermally insulating wall separates the system from its surroundings. For the process to be natural, either (a) work must be done on the system at a finite rate, so that the internal energy of the system increases; the entropy of the system increases even though it is thermally insulated; or (b) the system must do work on the surroundings, which then suffer increase of entropy, as well as gaining energy from the system.

- An isentropic process is customarily defined as an idealized quasi-static reversible adiabatic process, of transfer of energy as work. Otherwise, for a constant-entropy process, if work is done irreversibly, heat transfer is necessary, so that the process is not adiabatic, and an accurate artificial control mechanism is necessary; such is therefore not an ordinary natural thermodynamic process.

Chemical Potential - particle Number

The processes just above have assumed that the boundaries are also impermeable to particles. Otherwise, we may assume boundaries that are rigid, but are permeable to one or more types of particle. Similar considerations then hold for the chemical potential–particle number conjugate pair, which is concerned with the transfer of energy via this transfer of particles.

- In a *constant chemical potential process* the system is *particle-transfer connected*, by a particle-permeable boundary, to a constant-μ reservoir.
- The conjugate here is a constant particle number process. These are the processes outlined just above. There is no energy added or subtracted from the system by particle transfer. The system is *particle-transfer-insulated* from its environment by a boundary that is impermeable to particles, but permissive of transfers of energy as work or heat. These processes are the ones by which thermodynamic work and heat are defined, and for them, the system is said to be closed.

Thermodynamic Potentials

Any of the thermodynamic potentials may be held constant during a process. For example:

- An isenthalpic process introduces no change in enthalpy in the system.

Polytropic Processes

A polytropic process is a thermodynamic process that obeys the relation:

$$PV^n = C,$$

where P is the pressure, V is volume, n is any real number (the "polytropic index"), and C

is a constant. This equation can be used to accurately characterize processes of certain systems, notably the compression or expansion of a gas, but in some cases, liquids and solids.

Processes Classified by the Second Law of Thermodynamics

According to Planck, one may think of three main classes of thermodynamic process: natural, fictively reversible, and impossible or unnatural.

Natural Process

Only natural processes occur in nature. For thermodynamics, a natural process is a transfer between systems that increases the sum of their entropies, and is irreversible. Natural processes may occur spontaneously, or may be triggered in a metastable or unstable system, as for example in the condensation of a supersaturated vapour.

Fictively Reversible Process

To describe the geometry of graphical surfaces that illustrate equilibrium relations between thermodynamic functions of state, one can fictively think of so-called "reversible processes". They are convenient theoretical objects that trace paths across graphical surfaces. They are called "processes" but do not describe naturally occurring processes, which are always irreversible. Because the points on the paths are points of thermodynamic equilibrium, it is customary to think of the "processes" described by the paths as fictively "reversible". Reversible processes are always quasistatic processes, but the converse is not always true.

Unnatural Process

Unnatural processes are logically conceivable but do not occur in nature. They would decrease the sum of the entropies if they occurred.

Quasistatic Process

A quasistatic process is an idealized or fictive model of a thermodynamic "process" considered in theoretical studies. It does not occur in physical reality. It may be imagined as happening infinitely slowly so that the system passes through a continuum of states that are infinitesimally close to equilibrium.

Reversible Process (Thermodynamics)

In thermodynamics, a reversible process is a process whose direction can be "reversed" by inducing infinitesimal changes to some property of the system via its surroundings, while not increasing entropy. Throughout the entire reversible process, the system is in thermodynamic equilibrium with its surroundings. Since it would take an infinite amount of time for the reversible process to finish, perfectly reversible processes are impossible. However, if the system undergoing the changes responds much faster than the applied change, the deviation from reversibility may be negligible. In a *reversible cycle*, a reversible process which is cyclic, the system and its surroundings will be returned to their original states if the forward cycle is followed by the reverse cycle.

Thermodynamic processes can be carried out in one of two ways: reversibly or irreversibly. Reversibility refers to performing a reaction continuously at equilibrium. In an ideal thermodynamically reversible process, the energy from work performed by or on the system would be maximized, and that from heat would be minimized; heat cannot fully be converted to work and will always be lost to some degree (to the surroundings). The phenomenon of maximized work and minimized heat can be visualized on a pressure-volume curve, as the area beneath the equilibrium curve, representing work done. In order to maximize work, one must follow the equilibrium curve closely.

Irreversible processes, on the other hand, are a result of straying away from the curve, therefore decreasing the amount of overall work done; an irreversible process can be described as a thermodynamic process that leaves equilibrium. When described in terms of pressure and volume, it occurs when the pressure or the volume of a system changes so dramatically and instantaneously that the other (pressure or volume in this case) does not have time to catch up. A classic example of irreversibility is allowing a certain volume of gas to be released into a vacuum. By releasing pressure on a sample and thus allowing it to occupy a large space, the system and surroundings are not in equilibrium during the expansion process and there is little work done. However, significant work will be required, with a corresponding amount of energy dissipated as heat flow to the environment, in order to reverse the process (compressing the gas back to its original volume and temperature).

An alternative definition of a *reversible process* is a process that, after it has taken place, can be reversed and, when reversed, causes no change in either the system or its surroundings. In thermodynamic terms, a process "taking place" would refer to its transition from its initial state to its final state.

Irreversibility

In an irreversible process, finite changes are made; therefore the system is not at equilibrium throughout the process. At the same point in an irreversible cycle, the system will be in the same state, but the surroundings are permanently changed after each cycle.

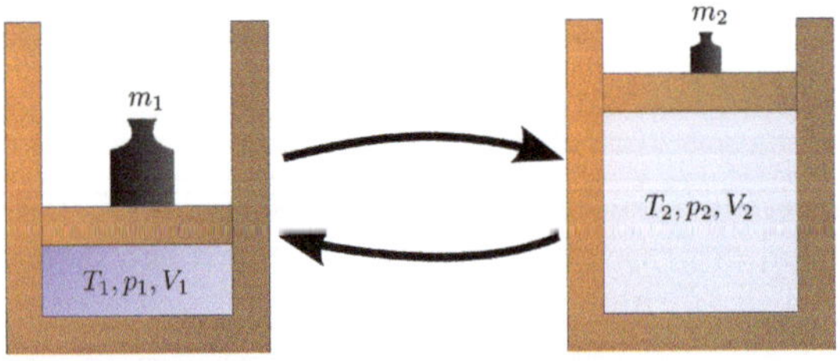

Reversible adiabatic process: The state on the left can be reached from the state on the right as well as vice versa without exchanging heat with the environment.

Boundaries and States

A reversible process changes the state of a system in such a way that the net change in the combined entropy of the system and its surroundings is zero. Reversible processes define the boundaries of how efficient heat engines can be in thermodynamics and engineering: a reversible process is one where no heat is lost from the system as "waste", and the machine is thus as efficient as it can possibly be.

In some cases, it is important to distinguish between reversible and quasistatic processes. Reversible processes are always quasistatic, but the converse is not always true. For example, an infinitesimal compression of a gas in a cylinder where there exists friction between the piston and the cylinder is a quasistatic, but not reversible process. Although the system has been driven from its equilibrium state by only an infinitesimal amount, heat has been irreversibly lost due to friction, and cannot be recovered by simply moving the piston infinitesimally in the opposite direction.

Engineering Archaisms

Historically, the term Tesla principle was used to describe (amongst other things) certain reversible processes invented by Nikola Tesla. However, this phrase is no longer in conventional use. The principle stated that some systems could be reversed and operated in a complementary manner. It was developed during Tesla's research in alternating currents where the current's magnitude and direction varied cyclically. During a demonstration of the Tesla turbine, the disks revolved and machinery fastened to the shaft was operated by the engine. If the turbine's operation was reversed, the disks acted as a pump.

- The objective is to build devices which receive heat and produce work (like an aircraft engine or a car engine) or receive work and produce heat (like an air conditioner) in a sustained manner.
- All operations need to be cyclic. The cycle comprises of a set of processes during which one of the properties is kept constant (V,p,T etc.)
- A minimum of 3 such processes are required to construct a cycle.
- All processes need not have work interactions (eg: isochoric)
- All processes need not involve heat interactions either (eg: adiabatic process).
- A cycle will consist of processes: involving some positive work interactions and some negative.
- If sum of +ve interactions is > -ve interactions the cycle will produce work
- If it is the other way, it will need work to operate.

- On the same lines some processes may have +ve and some -ve heat interactions.
- Commonsense tells us that to return to the same point after going round we need at one path of opposite direction.
- I law does not forbid all heat interactions being +ve nor all work interactions being -ve.
- But, we know that you can't construct a cycle with all +ve or
- All -ve Q's nor with all +ve or all -ve W's
- Any cycle you can construct will have some processes with
- Q +ve some with -ve.
- Let Q_1, Q_3, Q_5 .... be +ve heat interactions (Heat supplied)
- Q_2, Q_4, Q_6 .... be -ve heat interactions (heat rejected)
- From the first law we have
- $Q_1 + Q_3 + Q_5 .. - Q_2 - Q_4 - Q_6 - ...$ = Net work delivered (W_{net})
- $\Sigma Q_{+ve} - \Sigma Q_{-ve} = W_{net}$
- The efficiency of the cycle is defined as $\eta = W_{net} / \Sigma Q_{+ve}$
- Philosophy → What we have achieved ÷ what we have spent to achieve it

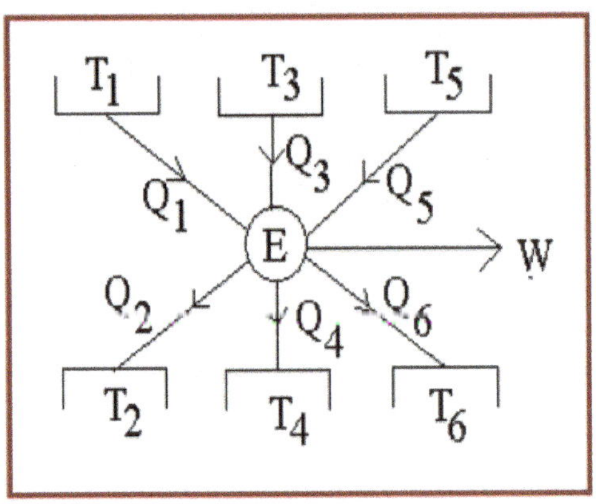

Otto Cycle

Consider the OTTO Cycle (on which your car engine works)

- There is no heat interaction during 1-2 and 3-4
- Heat is added during constant volume heating (2-3)Q2-3= cv (T3-T2)

- Heat is rejected during constant volume cooling (4-1) $Q_{4-1} = c_v(T_1 - T_4)$
- Which will be negative because $T_4 > T_1$

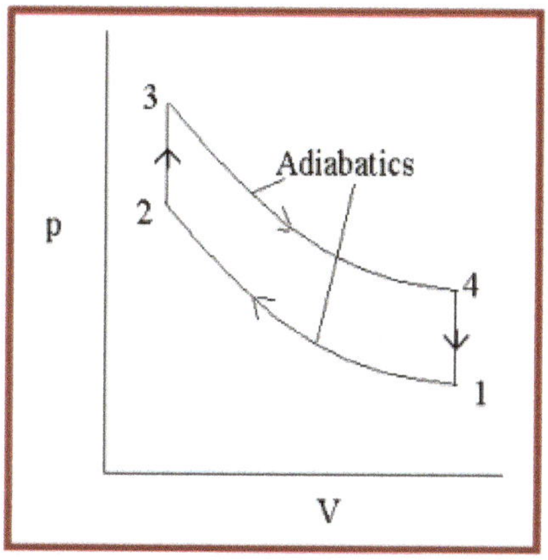

➢ Work done $= c_v(T3 - T2) + c_v(T1 - T4)$

➢ The efficiency $= [c_v(T3 - T2) + c_v(T1 - T4)]/[c_v(T3 - T2)]$

$$= [(T3 - T2) + (T1 - T4)]/[(T3 - T2)]$$

$$= 1 - [(T4 - T1)/(T3 - T2)]$$

Consider a Carnot cycle - against which all other cycles are compared

It consists of two isotherms and two adiabatics

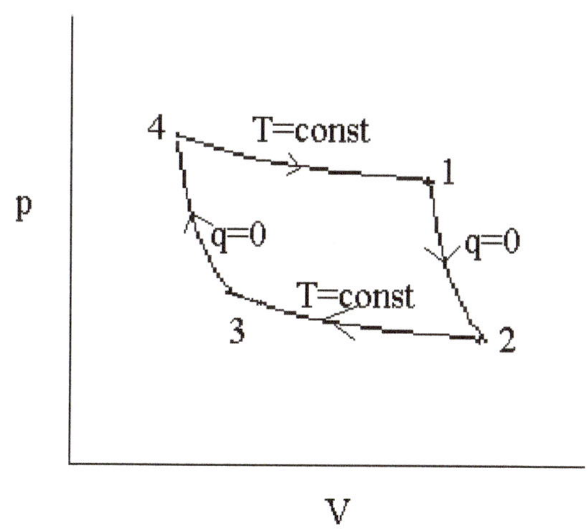

- Process 4-1 is heat addition because $v_4 < v_1$
- Process 2-3 is heat rejection because $v_3 < v_2$

Process	Work	Heat
1-2	$(p_1v_1-p_2v_2)/(g-1)$	0
2-3	$p_2v_2 \ln (v_3/v_2)$	$p_2v_2 \ln (v_3/v_2)$
3-4	$(p_3v_3-p_4v_4)/(g-1)$	0
4-1	$p_4v_4 \ln (v_1/v_4)$	$p_4v_4 \ln (v_1/v_4)$

Sum $(p_1v_1-p_2v_2 + p_3v_3-p_4v_4)/(g-1)$

$+ RT_2 \ln (v_3/v_2)$ $\qquad RT_2 \ln (v_3/v_2)$

$+ RT_1 \ln (v_1/v_4)$ $\qquad + RT_1 \ln (v_1/v_4)$

But, $p_1v_1 = p_4v_4$ and $p_2v_2 = p_3v_3$

Therefore the first term will be 0

!!We reconfirm that I law works!!

We will show that $(v_2/v_3) = (v_1/v_4)$

1 and 2 lie on an adiabatic so do 3 and 4

$p_1v_1^g = p_2v_2^g$ $\qquad p_4v_4^g = p_3v_3^g$

Divide one by the other $\qquad (p_1v_1^g / p_4v_4^g) = (p_2v_2^g / p_3v_3^g)$ $\qquad$ (A)

$$(p_1/p_4)(v_1^g / v_4^g) = (p_2/p_3)(v_2^g / v_3^g)$$

But $\quad (p_1/p_4) = (v_4/v_1)$ $\quad$ because 1 and 4 are on the same isotherm

Similarly $(p_2/p_3) = (v_3/v_2)$ because 2 and 3 are on the same isotherm

Therefore A becomes $\qquad (v_1/v_4)^{g-1} = (v_2/v_3)^{g-1}$

which means $\qquad (v_2/v_3) = (v_1/v_4)$

Work done in Carnot cycle $= RT_1 \ln (v_1/v_4) + RT_2 \ln (v_3/v_2)$

$$= RT_1 \ln (v_1/v_4) - RT_2 \ln (v_2/v_3)$$

$$= R \ln (v_1/v_4)(T_1 - T_2)$$

Heat supplied $= R \ln (v_1/v_4) T_1$

The efficiency $= (T_1 - T_2)/T_1$

In all the cycles it also follows that Work done=Heat supplied - heat rejected

Carnot engine has one Q +ve process and one Q -ve process.This engine has a single heat source at T_1 and a single sink at T_2.

If Q +ve > Q -ve; W will be +ve It is a heat engine

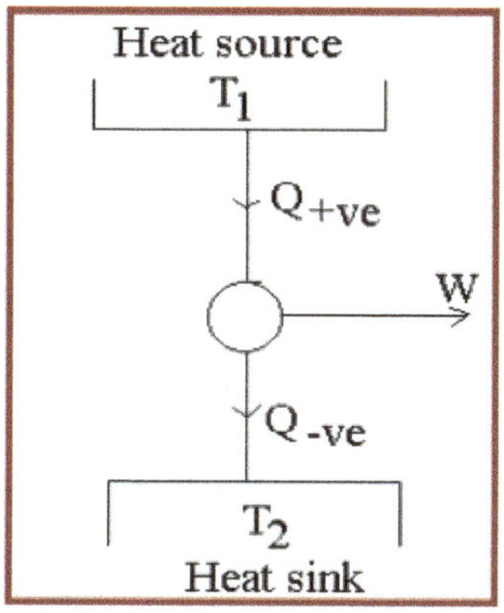

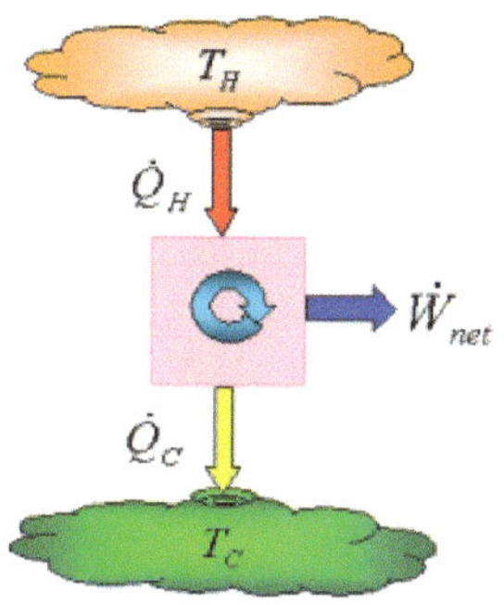

Fig. 1. Heat Engine

It will turn out that Carnot efficiency of $(T_1 - T_2)/T_1$ is the best we can get for any cycle operating between two fixed temperatures.

Q +ve < Q -ve W will be - ve It is not a heat engine

Efficiency is defined only for a work producing heat engine not a work consuming cycle

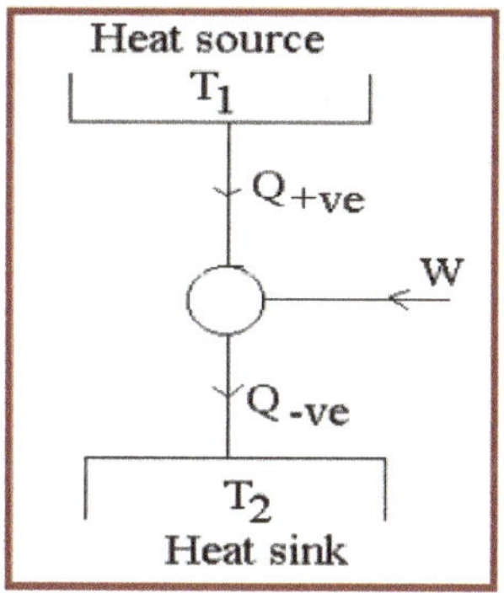

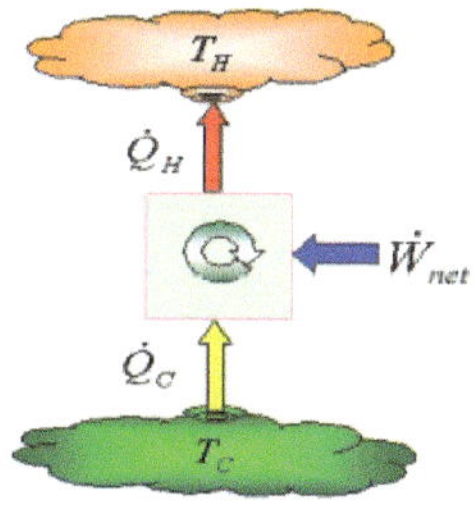

Fig. 1. Refrigerator/Heat Pump.

Note: We can't draw such a diagram for an Otto cycle because there is no single temperature at which heat interactions occur.

5

An Integrated Study of Thermodynamical Relations

Thermodynamics studies the conversion of energy from one form to another. The laws of thermodynamics are applied to almost every stream in science and technology. The topic can be classified into classical thermodynamics, statistical thermodynamics, kinetic theory and irreversible thermodynamics. The aspects elucidated in this chapter are of vital importance, and provide a better understanding of thermodynamics.

Thermodynamic Relations

In thermodynamics, an important task is to develop relationships which express the nonmeasurable properties like U,H,A,G and S in terms of measurable properties P,V & T The number of partial derivatives of the type $\left(\frac{\partial S}{\partial P}\right)_v$ among the above eight properties are $8C_3 \times 3! = 336$

If there exists a relation S=S (T,P) among the three variables S,T,P then

$$ds = \left(\frac{\partial S}{\partial T}\right)_P dT + \left(\frac{\partial S}{\partial P}\right)_T dP$$

Dividing the above relation with dT while V is held constant, we obtain

$$\left(\frac{\partial S}{\partial T}\right)_V = \left(\frac{\partial S}{\partial P}\right)_P + \left(\frac{\partial S}{\partial P}\right)_T \left(\frac{\partial P}{\partial T}\right)_V$$

There exists one relationship among four partial derivatives

$$\left(\frac{\partial S}{\partial T}\right)_V, \left(\frac{\partial S}{\partial T}\right)_P, \left(\frac{\partial S}{\partial P}\right)_T \text{ and } \left(\frac{\partial P}{\partial T}\right)_V$$

Therefore a large number of relationships is possible among 336 partial derivatives.

Important Mathematical Relations

Suppose there exists a relationship *f(x, y, z)=0* among the there variables *x, y* and z. Then the following relation can be obtained.

- The chain rule of partial differentiation

$$\left(\frac{\partial x}{\partial y}\right)_z = \left(\frac{\partial x}{\partial a}\right)_z \left(\frac{\partial a}{\partial y}\right)_z$$

- Reciprocal relation,

$$\left(\frac{\partial x}{\partial y}\right)_z = \frac{1}{\left(\frac{\partial y}{\partial x}\right)_z}$$

- Cyclic relation,

$$\left(\frac{\partial x}{\partial y}\right)_z \left(\frac{\partial y}{\partial z}\right)_x \left(\frac{\partial z}{\partial x}\right)_y = -1$$

Jacobian Methods

$$x = x(a,b)$$

$$x = y(a,b)$$

$$J\frac{\partial(x,y)}{\partial(\mathrm{a,b})} = \begin{vmatrix} \left(\frac{\partial x}{\partial a}\right)_b & \left(\frac{\partial x}{\partial b}\right)_a \\ \left(\frac{\partial y}{\partial a}\right)_b & \left(\frac{\partial y}{\partial b}\right)_a \end{vmatrix} = \left(\frac{\partial x}{\partial a}\right)_b \left(\frac{\partial y}{\partial b}\right)_a - \left(\frac{\partial x}{\partial b}\right)_a \left(\frac{\partial y}{\partial a}\right)_b$$

In general $J\left[\frac{x,y}{\mathrm{a,b}}\right]$ can be written as $\left[\frac{x,y}{\mathrm{a,b}}\right]$

Let $x = x(\mathrm{a,y})$

$y = y(\mathrm{a,y})$

This is a special case $(\mathrm{b,y})$

$$J\frac{\partial(x,y)}{\partial(\mathrm{a,y})} = \left(\frac{\partial x}{\partial a}\right)_y = \frac{[x,y]}{[\mathrm{a,y}]}$$

Similarly

$$\frac{[\mathrm{f,g}]}{[\mathrm{h,g}]} = \left(\frac{\partial f}{\partial h}\right)_g$$

1. Properties:

$$[a,a] = 0$$

2. Transposition

$$\frac{[x,y]}{[f,g]}=-\frac{[y,x]}{[f,g]}=-\frac{[x,y]}{[g,f]}=\frac{[y,x]}{[g,f]}$$

3. Inversion

$$\frac{[x,y]}{[f,g]}=\frac{1}{\frac{[f,g]}{[x,y]}}$$

4. Chain rule

$$\frac{[x,y]}{[f,g]}=\frac{[x,y]}{[a,b]}\cdot\frac{[a,b]}{[f,g]}$$

5. Cyclic rule

$$z=z(x,y)$$

$$dz=\left(\frac{\partial z}{\partial x}\right)_y dx+\left(\frac{\partial z}{\partial y}\right)_x dy$$

$$dz=\frac{[z,y]}{[x,y]}dx+\frac{[z,x]}{[y,x]_x}dy$$

$$dz=M\ \ dx+N^{dy}$$

$$[x,y]dz=[z,y]dx-[z,x]dy$$

or

$$[x,y]dz=[y,z]dx+[z,x]dy=0$$

Let us consider

$$x=x(a,b),\ y=y(a,b)$$

$$[x,y]dz=[z,y]dx-[z,x]dy$$

$$[x,y]=\left(\frac{\partial z}{\partial b}\right)_a+[y,z]\left(\frac{\partial x}{\partial b}\right)_a+[z,x]\left(\frac{\partial y}{\partial b}\right)_a=0$$

In Jacobian form

$$[x,y][z,a]+[y,z][x,a]+[z,x][y,a]=0$$

Again

$$dz = \frac{[z,y]}{[x,y]}dx + \frac{[z,x]}{[y,x]}dy$$

$$dz = M\,dx + N\,dy$$

Let us consider $x = x(a,b)$, $y = y(a,b)$

$$\left(\frac{\partial z}{\partial b}\right)_a = M\left(\frac{\partial x}{\partial b}\right)_a + N\left(\frac{\partial y}{\partial b}\right)_a$$

$$[Z,a] = M[x,a] + N[y,a]$$

Earlier we have obtained the Tds equations. These can be rewritten as

$$du = Tds - Pdv$$

$$dn = Tds + vdP$$

In addition, we know the availability function and Gibbs function

$$da = -sdT - Pdv$$

$$dg = -sdT + vdP$$

These equations can be generated through Mnemonic diagram (figure)

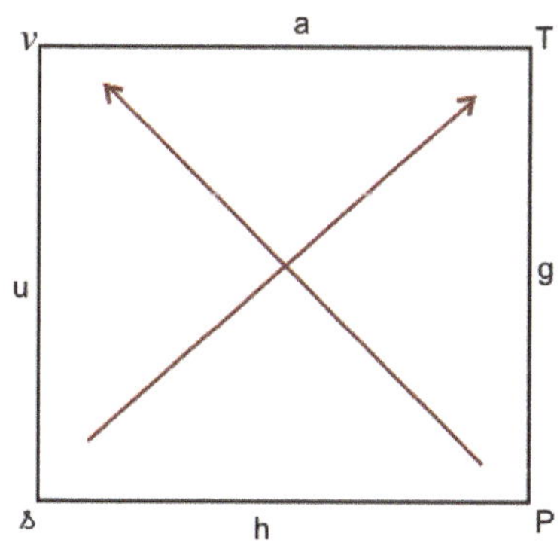

Differential (property) = differential of (independent variables)

$$du = (sign)(coeff)ds + (sign)(coeff)dv$$

$$dg = (sign)(coeff)dP + (sign)(coeff)dT$$

$$dh = (sign)(coeff)ds + (sign)(coeff)dP$$

$$da = (sign)(coeff)dv + (sign)(coeff)dT$$

Sign convention: going away from independent variable positive

Coming in toward independent variable negative

Equations (32.23)–(32.25) can now be expressed in the form

$$[Z,a] = M[\mathrm{x,a}] + N[\mathrm{y,a}]$$

Let b be the dummy variable

$$[u,b] = T[s,b] - P[\mathrm{v,b}]$$

$$[h,b] = T[s,b] + v[\mathrm{P,b}]$$

$$[\mathrm{a},b] = -s[\mathrm{T},b] - P[\mathrm{v,b}]$$

$$[\mathrm{g},b] = -s[\mathrm{T},b] + v[\mathrm{P,b}]$$

Maxwell Relations

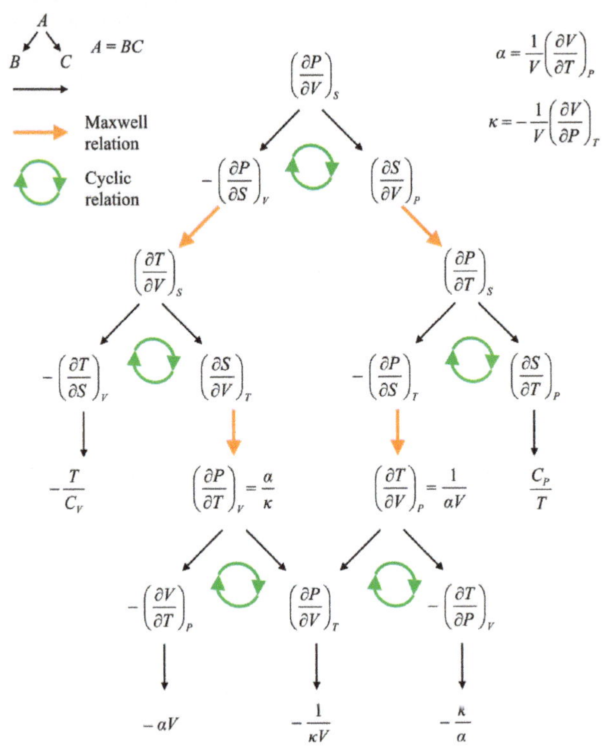

Flow chart showing the paths between the Maxwell relations. *P*: pressure, *T*: temperature, *V*: volume, *S*: entropy, α: coefficient of thermal expansion, κ: compressibility, C_V: heat capacity at constant volume, C_P: heat capacity at constant pressure.

Maxwell's relations are a set of equations in thermodynamics which are derivable from the symmetry of second derivatives and from the definitions of the thermodynamic potentials. These relations are named for the nineteenth-century physicist James Clerk Maxwell.

Equations

The structure of Maxwell relations is a statement of equality among the second derivatives for continuous functions. It follows directly from the fact that the order of differentiation of an analytic function of two variables is irrelevant (Schwarz theorem). In the case of Maxwell relations the function considered is a thermodynamic potential and x_i and x_j are two different natural variables for that potential:

Schwarz' Theorem (General)

$$\frac{\partial}{\partial x_j}\left(\frac{\partial \Phi}{\partial x_i}\right)=\frac{\partial}{\partial x_i}\left(\frac{\partial \Phi}{\partial x_j}\right)$$

where the partial derivatives are taken with all other natural variables held constant. It is seen that for every thermodynamic potential there are $n(n-1)/2$ possible Maxwell relations where n is the number of natural variables for that potential.

The four Most Common Maxwell Relations

The four most common Maxwell relations are the equalities of the second derivatives of each of the four thermodynamic potentials, with respect to their thermal natural variable (temperature *T*; or entropy *S*) and their *mechanical* natural variable (pressure *P*; or volume *V*):

Maxwell's Relations *(common)*

$$\begin{aligned}
+\left(\frac{\partial T}{\partial V}\right)_S &= -\left(\frac{\partial P}{\partial S}\right)_V = \frac{\partial^2 U}{\partial S \partial V}\\
+\left(\frac{\partial T}{\partial P}\right)_S &= +\left(\frac{\partial V}{\partial S}\right)_P = \frac{\partial^2 H}{\partial S \partial P}\\
+\left(\frac{\partial S}{\partial V}\right)_T &= +\left(\frac{\partial P}{\partial T}\right)_V = -\frac{\partial^2 F}{\partial T \partial V}\\
-\left(\frac{\partial S}{\partial P}\right)_T &= +\left(\frac{\partial V}{\partial T}\right)_P = \frac{\partial^2 G}{\partial T \partial P}
\end{aligned}$$

where the potentials as functions of their natural thermal and mechanical variables are the internal energy *U*(*S*, *V*), enthalpy *H*(*S*, *P*), Helmholtz free energy *F*(*T*, *V*) and Gibbs free energy *G*(*T*, *P*). The thermodynamic square can be used as a mnemonic to recall and derive these relations. The usefulness of these relations lies in their quantifying entropy changes, which are not directly measurable, in terms of measurable quantities like temperature, volume, and pressure.

Derivation

Maxwell relations are based on simple partial differentiation rules, in particular the total differential of a function and the symmetry of evaluating second order partial derivatives.

Derivation of the Maxwell relation can be deduced from the differential forms of the thermodynamic potentials:

The differential form of internal energy U is

$$dU = TdS - PdV$$

This equation resembles total differentials of the form

$$dz = \left(\frac{\partial z}{\partial x}\right)_y dx + \left(\frac{\partial z}{\partial y}\right)_x dy$$

It can be shown that for any equation of the form

$$dz = Mdx + Ndy$$

that

$$M = \left(\frac{\partial z}{\partial x}\right)_y, \quad N = \left(\frac{\partial z}{\partial y}\right)_x$$

Consider, the equation $dU = TdS - PdV$.. We can now immediately see that

$$T = \left(\frac{\partial U}{\partial S}\right)_V, \quad -P = \left(\frac{\partial U}{\partial V}\right)_S$$

Since we also know that for functions with continuous second derivatives, the mixed partial derivatives are identical (Symmetry of second derivatives), that is, that

$$\frac{\partial}{\partial y}\left(\frac{\partial z}{\partial x}\right)_y = \frac{\partial}{\partial x}\left(\frac{\partial z}{\partial y}\right)_x = \frac{\partial^2 z}{\partial y \partial x} = \frac{\partial^2 z}{\partial x \partial y}$$

we therefore can see that

$$\frac{\partial}{\partial V}\left(\frac{\partial U}{\partial S}\right)_V = \frac{\partial}{\partial S}\left(\frac{\partial U}{\partial V}\right)_S$$

and therefore that

$$\left(\frac{\partial T}{\partial V}\right)_S = -\left(\frac{\partial P}{\partial S}\right)_V$$

Derivation of Maxwell Relation from Helmholtz Free energy

The differential form of Helmholtz free energy is

$$dF = -SdT - PdV$$

$$-S = \left(\frac{\partial F}{\partial T}\right)_V, \quad -P = \left(\frac{\partial F}{\partial V}\right)_T$$

From symmetry of second derivatives

$$\frac{\partial}{\partial V}\left(\frac{\partial F}{\partial T}\right)_V = \frac{\partial}{\partial T}\left(\frac{\partial F}{\partial V}\right)_T$$

and therefore that

$$\left(\frac{\partial S}{\partial V}\right)_T = \left(\frac{\partial P}{\partial T}\right)_V$$

The other two Maxwell relations can be derived from differential form of enthalpy $dH = TdS + VdP$ and the differential form of Gibbs free energy $dG = VdP - SdT$ in a similar way. So all Maxwell Relationships above follow from one of the Gibbs equations.

Derivation Based on Jacobians

If we view the first law of thermodynamics,

$$dU = TdS - PdV$$

as a statement about differential forms, and take the exterior derivative of this equation, we get

$$0 = dTdS - dPdV$$

since $d(dU) = 0.$. This leads to the fundamental identity

$$dPdV = dTdS.$$

The physical meaning of this identity can be seen by noting that the two sides are the equivalent ways of writing the work done in an infinitesimal Carnot cycle. An equivalent way of writing the identity is

$$\frac{\partial(T,S)}{\partial(P,V)} = 1.$$

The Maxwell relations now follow directly. For example,

$$\left(\frac{\partial S}{\partial V}\right)_T = \frac{\partial(T,S)}{\partial(T,V)} = \frac{\partial(P,V)}{\partial(T,V)} = \left(\frac{\partial P}{\partial T}\right)_V,$$

et Voila! The critical step is the penultimate one. The other Maxwell relations follow in similar fashion. For example,

$$\left(\frac{\partial T}{\partial V}\right)_S = \frac{\partial(T,S)}{\partial(V,S)} = \frac{\partial(P,V)}{\partial(V,S)} = -\left(\frac{\partial P}{\partial S}\right)_V.$$

General Maxwell Relationships

The above are not the only Maxwell relationships. When other work terms involving other natural variables besides the volume work are considered or when the number of particles is included as a natural variable, other Maxwell relations become apparent. For example, if we have a single-component gas, then the number of particles N is also a natural variable of the above four thermodynamic potentials. The Maxwell relationship for the enthalpy with respect to pressure and particle number would then be:

$$\left(\frac{\partial \mu}{\partial P}\right)_{S,N} = \left(\frac{\partial V}{\partial N}\right)_{S,P} = \frac{\partial^2 H}{\partial P \partial N}$$

where μ is the chemical potential. In addition, there are other thermodynamic potentials besides the four that are commonly used, and each of these potentials will yield a set of Maxwell relations.

Each equation can be re-expressed using the relationship

$$\left(\frac{\partial y}{\partial x}\right)_z = 1/\left(\frac{\partial x}{\partial y}\right)_z$$

which are sometimes also known as Maxwell relations.

Maxwell relations are links between the independent variables. We require focus on the independent variables and the directions of the arrows only.

From figure we can write

$$\left(\frac{\partial v}{\partial s}\right)_P = \left(\frac{\partial T}{\partial P}\right)_s$$

Since both the arrow are in the same direction (upward), there is no change in sign. For the next Maxwell relation, we require. To rotate the Mnemonic diagram by 90° (Figure)

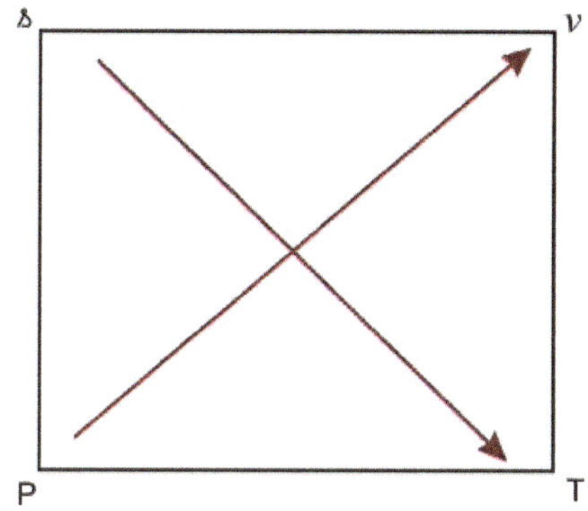

$$\left(\frac{\partial s}{\partial P}\right)_T = -\left(\frac{\partial u}{\partial T}\right)_P$$

For next relation, we require to rotate it by 90° again (Figure)

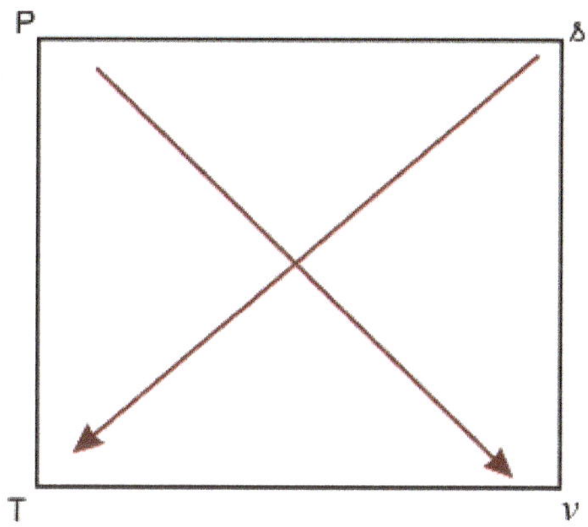

$$\left(\frac{\partial P}{\partial T}\right)_v = -\left(\frac{\partial s}{\partial v}\right)_T$$

Similarly for the other relation we require to rotate another 90° (Figure)

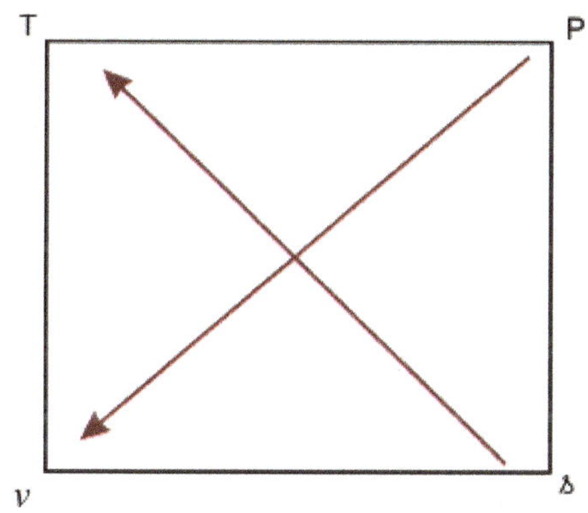

And the relation is

$$-\left(\frac{\partial T}{\partial v}\right)_s = \left(\frac{\partial P}{\partial s}\right)_v$$

Equations can be re-written as

$$\frac{[v,P]}{[s,P]} = \frac{[T,s]}{[P,s]}$$

$$\frac{[s,T]}{[P,T]} = -\frac{[v,P]}{[T,P]}$$

$$\frac{[P,v]}{[T,v]} = -\frac{[s,T]}{[v,T]}$$

$$-\frac{[T,s]}{[v,s]} = \frac{[P,v]}{[s,v]}$$

All four relations can be summarized as

or $\left.\begin{matrix}[s,T]=[v,P] \\ [T,s]=[P,v]\end{matrix}\right\}$

Other measurable quantities, such as

Coefficient of volume expansion, $\beta = \frac{1}{v}\left(\frac{\partial v}{\partial T}\right)_P = \frac{1}{v}\frac{[v,P]}{[T,P]}$

Compressibility $k = -\frac{1}{v}\left(\frac{\partial v}{\partial P}\right)_T = -\frac{1}{v}\frac{[v,T]}{[P,T]}$

Specific heat at constant pressure $c_p = T\left(\frac{\partial s}{\partial T}\right)_P = T\frac{[s,P]}{[T,P]}$

Specific heat at constant volume $c_v = T\left(\frac{\partial s}{\partial T}\right)_v = T\frac{[s,v]}{[T,v]}$

Therfore $\frac{\beta}{k} = \frac{-[v,P]}{[T,P]}\frac{[P,T]}{[v,T]} = \frac{[v,P]}{[v,T]}$

Examples: Thermodynamics Relations Involving Entropy

The entropy of a substance can be expressed in terms of independent variable (T,P) or (T,v) and these relations are quite useful in estimating the entropy change when the system changes its state from (P_1,v_1,T_1) to (P_2,v_2,T_2)

(a) Entropy as a function of T and P

$$s = s(T,P)$$

$$ds = \left(\frac{\partial s}{\partial T}\right)_P dT + \left(\frac{\partial s}{\partial P}\right)_T dP$$

or

$$ds = \frac{[s,P]}{[T,P]}dT + \frac{[s,T]}{[P,T]}dp$$

Maxwell relation $[T,s]=[P,v]$ and $c_P = \frac{T[s,P]}{[T,P]}$

$$ds = \frac{c_P}{T}dT - \frac{[P,v]}{[P,T]}dP$$

$$ds = \frac{c_P}{T}dT - \beta v dP$$

or

$$Tds = c_P dT - \beta vTdP$$

(b) Entropy a as function of T and v

$$ds = \left(\frac{\partial s}{\partial T}\right)_v dT + \left(\frac{\partial s}{\partial v}\right)_T dv$$

$$ds = \frac{[s,v]}{[T,v]}dT + \frac{[s,T]}{[v,T]}dv$$

$$ds = \frac{c_v}{T}dT + \frac{[v,P]}{[v,T]}dv$$

using

$$[T,s]=[P,v]$$

$$ds = \frac{c_v}{T}dT + \frac{\beta}{k}dv$$

or

$$Tds = c_v dT + \frac{T\beta}{k}dv$$

Clausius–Clapeyron Relation

The Clausius–Clapeyron relation, named after Rudolf Clausius and Benoît Paul Émile Clapeyron, is a way of characterizing a discontinuous phase transition between two phases of matter of a single constituent. On a pressure–temperature (P–T) diagram, the line separating the two phases is known as the coexistence curve. The Clausius–Clapeyron relation gives the slope of the tangents to this curve. Mathematically,

$$\frac{dP}{dT} = \frac{L}{T\Delta v} = \frac{\Delta s}{\Delta v},$$

where dP/dT is the slope of the tangent to the coexistence curve at any point, L is the specific latent heat, T is the temperature, Δv is the specific volume change of the phase transition, and Δs is the specific entropy change of the phase transition.

Derivations

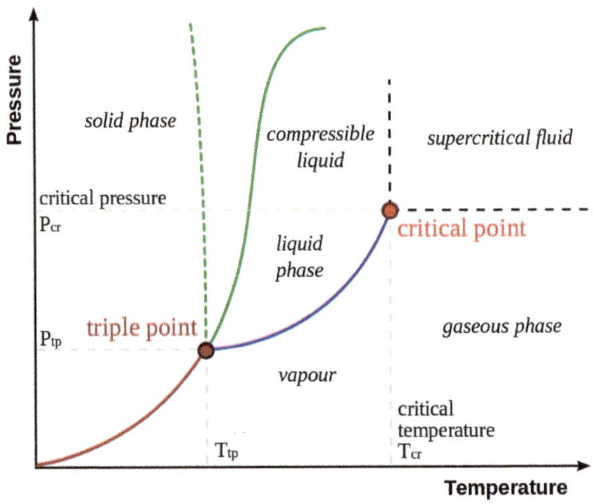

A typical phase diagram. The dotted green line gives the anomalous behavior of water. The Clausius–Clapeyron relation can be used to find the relationship between pressure and temperature along phase boundaries.

Derivation from State Postulate

Using the state postulate, take the specific entropy s for a homogeneous substance to be a function of specific volume v and temperature T.

$$ds = \left(\frac{\partial s}{\partial v}\right)_T dv + \left(\frac{\partial s}{\partial T}\right)_v dT$$

The Clausius–Clapeyron relation characterizes behavior of a closed system during a phase change, during which temperature and pressure are constant by definition. Therefore,

$$ds = \left(\frac{\partial s}{\partial v}\right)_T dv$$

Using the appropriate Maxwell relation gives

$$ds = \left(\frac{\partial P}{\partial T}\right)_v dv$$

where P is the pressure. Since pressure and temperature are constant, by definition the derivative of pressure with respect to temperature does not change. Therefore, the partial derivative of specific entropy may be changed into a total derivative

$$ds = \frac{dP}{dT} dv$$

and the total derivative of pressure with respect to temperature may be factored out when integrating from an initial phase α to a final phase β, to obtain

$$\frac{dP}{dT} = \frac{\Delta s}{\Delta v}$$

where $\Delta s \equiv s_\beta - s_\alpha$ and $\Delta v \equiv v_\beta - v_\alpha$ are respectively the change in specific entropy and specific volume. Given that a phase change is an internally reversible process, and that our system is closed, the first law of thermodynamics holds

$$du = \delta q + \delta w = T\, ds - P\, dv$$

where u is the internal energy of the system. Given constant pressure and temperature (during a phase change) and the definition of specific enthalpy h, we obtain

$$dh = du + P\, dv$$

$$dh = T\, ds$$

$$ds = \frac{dh}{T}$$

Given constant pressure and temperature (during a phase change), we obtain

$$\Delta s = \frac{\Delta h}{T}$$

Substituting the definition of specific latent heat $L = \Delta h$ gives

$$\Delta s = \frac{L}{T}$$

Substituting this result into the pressure derivative given above ($dP / dT = \Delta s / \Delta v$), we obtain

$$\frac{dP}{dT} = \frac{L}{T \Delta v}.$$

This result (also known as the Clapeyron equation) equates the slope of the tangent to the coexistence curve dP / dT, at any given point on the curve, to the function $L / T \Delta v$ of the specific latent heat L, the temperature T, and the change in specific volume Δv .

Derivation from Gibbs–Duhem Relation

Suppose two phases, α and β, are in contact and at equilibrium with each other. Their chemical potentials are related by

$$\mu_\alpha = \mu_\beta.$$

Furthermore, along the coexistence curve,

$$\mathrm{d}\mu_\alpha = \mathrm{d}\mu_\beta.$$

One may therefore use the Gibbs–Duhem relation

$$\mathrm{d}\mu = M(-s\mathrm{d}T + v\mathrm{d}P)$$

(where s is the specific entropy, v is the specific volume, and M is the molar mass) to obtain

$$-(s_\beta - s_\alpha)\mathrm{d}T + (v_\beta - v_\alpha)\mathrm{d}P = 0$$

Rearrangement gives

$$\frac{\mathrm{d}P}{\mathrm{d}T} = \frac{s_\beta - s_\alpha}{v_\beta - v_\alpha} = \frac{\Delta s}{\Delta v}$$

from which the derivation of the Clapeyron equation continues as in the previous section.

Ideal Gas Approximation at Low Temperatures

When the phase transition of a substance is between a gas phase and a condensed phase (liquid or solid), and occurs at temperatures much lower than the critical temperature of that substance, the specific volume of the gas phase v_g greatly exceeds that of the condensed phase v_c. Therefore, one may approximate

$$\Delta v = v_\mathrm{g}\left(1 - \frac{v_\mathrm{c}}{v_\mathrm{g}}\right) \approx v_\mathrm{g}$$

at low temperatures. If pressure is also low, the gas may be approximated by the ideal gas law, so that

$$v_\mathrm{g} = RT / P$$

where P is the pressure, R is the specific gas constant, and T is the temperature. Substituting into the Clapeyron equation

$$\frac{\mathrm{d}P}{\mathrm{d}T} = \frac{\Delta s}{\Delta v}$$

we can obtain the Clausius–Clapeyron equation

$$\frac{\mathrm{d}P}{\mathrm{d}T}=\frac{PL}{T^2R}$$

for low temperatures and pressures, where L is the specific latent heat of the substance.

Let (P_1,T_1) and (P_2,T_2) be any two points along the coexistence curve between two phases α and β. In general, L varies between any two such points, as a function of temperature. But if L is constant,

$$\frac{\mathrm{d}P}{P}=\frac{L}{R}\frac{\mathrm{d}T}{T^2},$$

$$\int_{P_1}^{P_2}\frac{\mathrm{d}P}{P}=\frac{L}{R}\int_{T_1}^{T_2}\frac{\mathrm{d}T}{T^2}$$

$$\ln P\Big|_{P=P_1}^{P_2}=-\frac{L}{R}\cdot\frac{1}{T}\Big|_{T=T_1}^{T_2}$$

or

$$\ln\frac{P_1}{P_2}=-\frac{L}{R}\left(\frac{1}{T_1}-\frac{1}{T_2}\right)$$

These last equations are useful because they relate equilibrium or saturation vapor pressure and temperature to the latent heat of the phase change, *without* requiring specific volume data.

Applications

Chemistry and Chemical Engineering

For transitions between a gas and a condensed phase with the approximations described above, the expression may be rewritten as

$$\ln P=-\frac{L}{R}\left(\frac{1}{T}\right)+c$$

where c is a constant. For a liquid-gas transition, L is the specific latent heat (or specific enthalpy) of vaporization; for a solid-gas transition, L is the specific latent heat of sublimation. If the latent heat is known, then knowledge of one point on the coexistence curve determines the rest of the curve. Conversely, the relationship between $\ln P$ and $1/T$ is linear, and so linear regression is used to estimate the latent heat.

Meteorology and Climatology

Atmospheric water vapor drives many important meteorologic phenomena (notably precipitation), motivating interest in its dynamics. The Clausius–Clapeyron equation for water vapor under typical atmospheric conditions (near standard temperature and pressure) is

$$\frac{de_s}{dT} = \frac{L_v(T)e_s}{R_v T^2}$$

where:

- e_s is saturation vapor pressure
- T is temperature
- L_v is the specific latent heat of evaporation of water
- R_v is the gas constant of water vapor

The temperature dependence of the latent heat $L_v(T)$, and therefore of the saturation vapor pressure $e_s(T)$, cannot be neglected in this application. Fortunately, the August-Roche-Magnus formula provides a very good approximation, using pressure in hPa and temperature in Celsius:

$$e_s(T) = 6.1094 \exp\left(\frac{17.625T}{T + 243.04}\right)$$

(This is also sometimes called the *Magnus* or *Magnus-Tetens* approximation, though this attribution is historically inaccurate.)

Under typical atmospheric conditions, the denominator of the exponent depends weakly on T (for which the unit is Celsius). Therefore, the August-Roche-Magnus equation implies that saturation water vapor pressure changes approximately exponentially with temperature under typical atmospheric conditions, and hence the water-holding capacity of the atmosphere increases by about 7% for every 1 °C rise in temperature.

Example

One of the uses of this equation is to determine if a phase transition will occur in a given situation. Consider the question of how much pressure is needed to melt ice at a temperature ΔT below 0 °C. Note that water is unusual in that its change in volume upon melting is negative. We can assume

$$\Delta P = \frac{L}{T \Delta v} \Delta T$$

and substituting in

$L = 3.34 \times 10^5$ J/kg (latent heat of fusion for water),

T = 273 K (absolute temperature), and

$\Delta v = -9.05\times10^{-5}$ m³/kg (change in specific volume from solid to liquid),

we obtain

$$\frac{\Delta P}{\Delta T} = -13.5 \text{ MPa/K}.$$

To provide a rough example of how much pressure this is, to melt ice at −7 °C (the temperature many ice skating rinks are set at) would require balancing a small car (mass = 1000 kg) on a thimble (area = 1 cm²).

Second Derivative

While the Clausius–Clapeyron relation gives the slope of the coexistence curve, it does not provide any information about its curvature or second derivative. The second derivative of the coexistence curve of phases 1 and 2 is given by

$$\frac{d^2P}{dT^2} = \frac{1}{v_2 - v_1}\left[\frac{c_{p2} - c_{p1}}{T} - 2(v_2\alpha_2 - v_1\alpha_1)\frac{dP}{dT}\right] + \frac{1}{v_2 - v_1}\left[(v_2\kappa_{T2} - v_1\kappa_{T1})\left(\frac{dP}{dT}\right)^2\right],$$

where subscripts 1 and 2 denote the different phases, c_p is the specific heat capacity at constant pressure, $\alpha = (1/v)(dv/dT)_P$ is the thermal expansion coefficient, and $\kappa_T = -(1/v)(dv/dP)_T$ is the isothermal compressibility.

Clapeyron Equations

Let us study v–T diagram of a system. The system consists of a liquid phase at state A and a vapour phase at state B (figure)

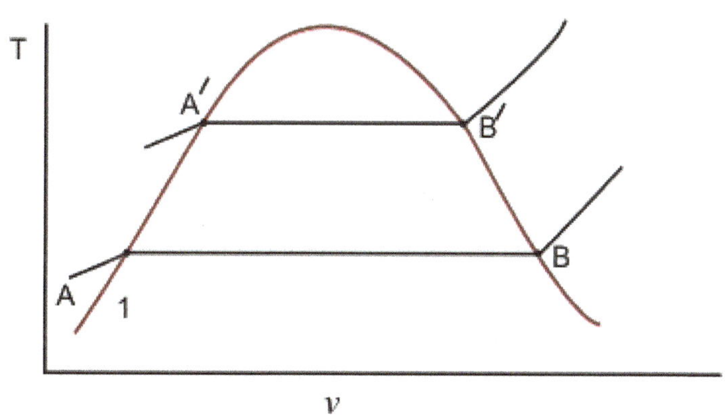

The line AB corresponds to the pressure P and temperature T. Gibbs free energy can be conveniently used in analyzing the problem. Using Gibbs equation we can write

$$dg = vdP - sdT$$

Along AB both dP and dT vanish; therefore dg=0

Hence, Gibbs function at a given pressure and temp has the same value for the saturated liquid and the saturated vapour.

The same result holds good for solid-liquid transition.

Next, consider a neighbouring line $A'B'$ corresponding to pressure $P+\Delta P$ and temperature $T+\Delta T$. The value of the specific Gibbs function on the line $A'B'$ may be denoted by $g+\Delta g$ Then assuming ΔT etc to be small, the change along AA' is given by

$$\Delta g = v_f \Delta P - s_f \Delta T$$

While the same along BB' is given by

$$\Delta g = v_g \Delta P - s_g \Delta T$$

Hence form we get,

$$\left(v_g - v_f\right)\Delta P = \left(s_g - s_f\right)\Delta T$$

Diving by ΔT and proceeding to the limit $\Delta T \to 0$

$$\left(\frac{\partial P}{\partial T}\right)_{sat} = \frac{s_g - s_f}{v_g - v_f}$$

As transition from A to B is a constant pressure and constant temperature process, we get

$$s_g - s_f = \frac{h_g - h_f}{T} = \frac{L}{T} = \frac{h_{fg}}{T}$$

Where L or is the latent heat of evaporation at temperature T

Hence

$$\left(\frac{\partial P}{\partial T}\right)_{sat} = \frac{h_{fg}}{T\left(v_g - v_f\right)} = \frac{h_{fg}}{T\left(\hat{v}_g - \hat{v}_f\right)}$$

This is known as Clapeyron Equation. Also valid for solid - liquid transition.

- h_{fg} is always positive so $\left(\frac{\partial P}{\partial T}\right)$ will depend on $\left(v_g - v_f\right)$
- Liquid $\to$ vapour: $\left(v_g - v_f\right)$ is always positive.

At higher pressure, boiling point is higher.

- But for solid → liquid: sometimes $(v_f - v_s)$ is negative; Notably for water: Melting is accompanied by contraction. A rise is pressure will lower down the melting point. (figure) This explains the phenomenon of relegation.

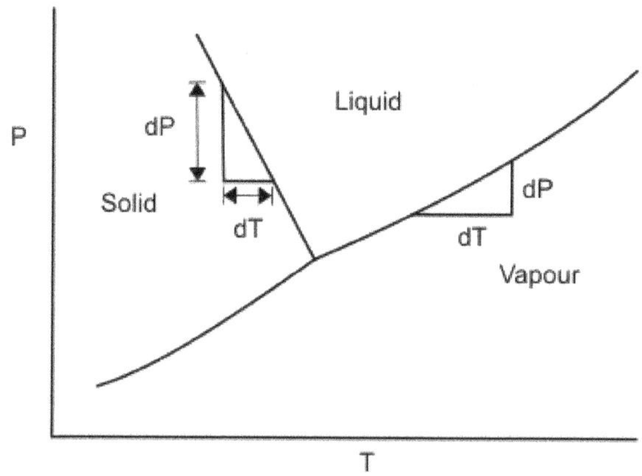

The Clapeyron equation can be put in a simple form if we make certain approximations. Let us consider the liquid – vapour phase transition at low pressures. Then the vapour phase may be approximated as an ideal gas. Moreover the volume if the liquid phase is negligible compared to the volume of the vapour phase $(v_g \gg v_f)$.

Hence

$$v_g - v_f \approx \Delta v \approx v_g = \frac{RT}{P}$$

Using these condition, Clapeyron equation becomes

$$\left(\frac{\partial P}{\partial T}\right)_{sat} = \frac{h_{fg}P}{RT^2}$$

or

$$\frac{d(InP)}{dT} = \frac{h_{fg}}{RT^2}$$

Which is known as Clausius – Clapeyron equation further, if we assume that h_{fs} is constant over a small temperature range, the above equation may be integrated to get

$$In\frac{P_2}{P_1} = \frac{h_{fg}}{R}\left(\frac{1}{T_1} - \frac{1}{T_2}\right)$$

or

$$In\,P = -\frac{h_{fg}}{RT} + Constant$$

Therefore a plot of InP versus 1/T yields a straight line the slope of which is equal to $\left(-h_{fg}/R\right)$. The vapour pressure of most the substances agree fairly well with this equation.

Change of Latent Heat with Temperature

The latent heat

$$h_{fg} = h_g - h_f$$

In going from AB to $A'B'$

$$\Delta h_{fg} = \Delta h_g - \Delta h_f$$

Again

$$dh = \left(\frac{\partial h}{\partial T}\right)_P dt + \left(\frac{\partial h}{\partial P}\right)_T dP$$

or

$$dh = c_p dt + \left\{v - T\left(\frac{\partial v}{\partial T}\right)_P\right\} dP$$

or

$$dh = c_p dt + v\left(1 - \beta T\right) dP$$

or

$$\Delta h_{fg} = \left(c_{Pg} - c_{Pf}\right) dT + \left\{\left(v_g - v_f\right) - \left(v_g \beta_g - v_f \beta_f\right) T\right\} dP$$

or

$$\left(\frac{\partial h_{fg}}{\partial T}\right) = \left(c_{Pg} - c_{Pf}\right) + \left\{\left(v_g - v_f\right) - \left(v_g \beta_g - v_f \beta_f\right) T\right\} \left(\frac{\partial P}{\partial T}\right)_{sat}$$

or

$$\frac{\partial h_{fg}}{\partial T} = \left(c_{Pg} - c_{Pf}\right) - \frac{\left(v_g \beta_g - v_f \beta_f\right) h_{fg}}{\left(v_g - v_f\right)} + \frac{h_{fg}}{T}$$

This is known as Kirchoff equation . This can be rewritten as:

$$\frac{\partial\left(h_{fg}/T\right)}{\partial T} + \frac{\left(v_g \beta_g - v_f \beta_f\right)}{\left(v_g - v_f\right)} \frac{h_{fg}}{T} = \frac{\left(c_{Pg} - c_{Pf}\right)}{T}$$

Two Special Cases:

a. In solid to liquid transition, the specific volume and specific heat are constant in each phase and the value of β in each phase is negligible; consequently,

$$\frac{\partial\left(h_{sf}/T\right)}{\partial T}=\frac{c_{p.f}-c_{P.s}}{T}$$

Where h_{sf} is the latent heat of Fusion

b. If the transition is from liquid to ideal gas, then $\beta_g=\frac{1}{T}$

Also, $v_g \gg v_f, \beta_g \gg \beta_f$ so that $v_g\beta_g \gg v_f\beta_f$

$$\frac{\partial h_{fg}}{\partial T}=c_{Pg}-c_{Pf}$$

This equation is also applicable in the case of sublimation with subscript f (for liquid) replaced by subscript s (for solid).

Phase Equilibrium

If two phases are in a state of equilibrium, the temperature and pressure of both the phases must be identical. The chemical potential μ_i of component i must be identical in both the phases. That is

$\mu_i^\alpha=\mu_i^\beta$ for all i

Where α and β refer to two different phases.

The criterion of equilibrium at constant temperature and pressure is given by $dG=0$

Phase Rule

Gibbs' phase rule was proposed by Josiah Willard Gibbs in his landmark paper titled *On the Equilibrium of Heterogeneous Substances*, published from 1875 to 1878. The rule applies to non-reactive multi-component heterogeneous systems in thermodynamic equilibrium and is given by the equality

$$F=C-P+2$$

where F is the number of degrees of freedom, C is the number of components and P is the number of phases in thermodynamic equilibrium with each other.

The number of degrees of freedom is the number of independent intensive variables, i.e. the largest number of thermodynamic parameters such as temperature or pressure that can be varied simultaneously and arbitrarily without affecting one another. An example of one-component system is a system involving one pure chemical, while two-component systems, such as mixtures of water and ethanol, have two chemically independent components, and so on. Typical phases are solids, liquids and gases.

Foundations

- A phase is a form of matter that is homogeneous in chemical composition and physical state. Typical phases are solid, liquid and gas. Two immiscible liquids (or liquid mixtures with different compositions) separated by a distinct boundary are counted as two different phases, as are two immiscible solids.
- The number of components (*C*) is the number of chemically independent constituents of the system, i.e. the minimum number of independent species necessary to define the composition of all phases of the system.
- The number of degrees of freedom (*F*) in this context is the number of intensive variables which are independent of each other.

The basis for the rule (Atkins and de Paula, justification 6.1) is that equilibrium between phases places a constraint on the intensive variables. More rigorously, since the phases are in thermodynamic equilibrium with each other, the chemical potentials of the phases must be equal. The number of equality relationships determines the number of degrees of freedom. For example, if the chemical potentials of a liquid and of its vapour depend on temperature (*T*) and pressure (*p*), the equality of chemical potentials will mean that each of those variables will be dependent on the other. Mathematically, the equation $\mu_{liq}(T, p) = \mu_{vap}(T, p)$, where μ = chemical potential, defines temperature as a function of pressure or vice versa. (Caution: do not confuse p = pressure with P = number of phases.)

To be more specific, the composition of each phase is determined by $C - 1$ intensive variables (such as mole fractions) in each phase. The total number of variables is $(C - 1)P + 2$, where the extra two are temperature T and pressure p. The number of constraints is $C(P - 1)$, since the chemical potential of each component must be equal in all phases. Subtract the number of constraints from the number of variables to obtain the number of degrees of freedom as $F = (C - 1)P + 2 - C(P - 1) = C - P + 2$.

The rule is valid provided the equilibrium between phases is not influenced by gravitational, electrical or magnetic forces, or by surface area, and only by temperature, pressure, and concentration.

Consequences and Examples

Pure Substances (One Component)

For pure substances $C = 1$ so that $F = 3 - P$. In a single phase ($P = 1$) condition of a pure component system, two variables ($F = 2$), such as temperature and pressure, can be chosen independently to be any pair of values consistent with the phase. However, if the temperature and pressure combination ranges to a point where the pure component undergoes a separation into two phases ($P = 2$), F decreases from 2 to 1. When the system enters the two-phase region, it becomes no longer possible to independently control temperature and pressure.

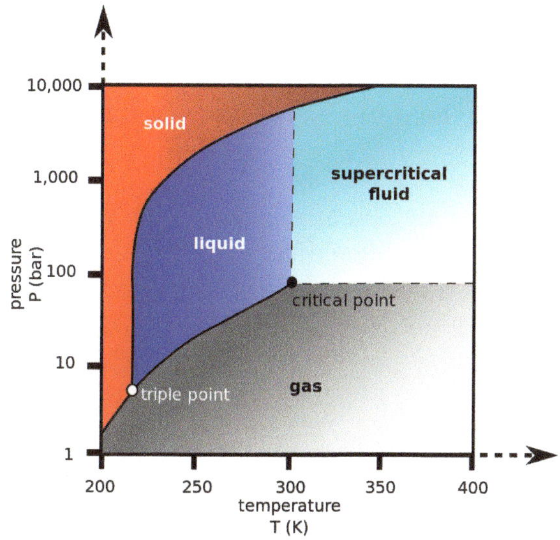

Carbon dioxide pressure-temperature phase diagram showing the triple point and critical point of carbon dioxide

In the phase diagram to the right, the boundary curve between the liquid and gas regions maps the constraint between temperature and pressure when the single-component system has separated into liquid and gas phases at equilibrium. If the pressure is increased by compression, some of the gas condenses and the temperature goes up. If the temperature is decreased by cooling, some of the gas condenses, decreasing the pressure. Throughout both processes, the temperature and pressure stay in the relationship shown by this boundary curve unless one phase is entirely consumed by evaporation or condensation, or unless the critical point is reached. As long as there are two phases, there is only one degree of freedom, which corresponds to the position along the phase boundary curve.

The critical point is the black dot at the end of the liquid–gas boundary. As this point is approached, the liquid and gas phases become progressively more similar until, at the critical point, there is no longer a separation into two phases. Above the critical point and away from the phase boundary curve, $F = 2$ and the temperature and pressure can be controlled independently. Hence there is only one phase, and it has the physical properties of a dense gas, but is also referred to as a supercritical fluid.

Of the other two-boundary curves, one is the solid–liquid boundary or melting point curve which indicates the conditions for equilibrium between these two phases, and the other at lower temperature and pressure is the solid–gas boundary.

Even for a pure substance, it is possible that three phases, such as solid, liquid and vapour, can exist together in equilibrium (P = 3). If there is only one component, there are no degrees of freedom (F = 0) when there are three phases. Therefore, in a single-component system, this three-phase mixture can only exist at a single temperature and pressure, which is known as a triple point. Here there are two equations $\mu_{sol}(T, p) = \mu_{liq}(T, p) = \mu_{vap}(T, p)$, which are sufficient to determine the two variables T and p. In the diagram for CO_2 the triple point is the point at which the solid, liquid and gas phases come together, at 5.2 bar and 217 K. It is also possible for other sets of phases to form a triple point, for example in the water system there is a triple point where ice I, ice III and liquid can coexist.

If four phases of a pure substance were in equilibrium (P = 4), the phase rule would give $F = -1$, which is meaningless, since there cannot be −1 independent variables. This explains the fact that four phases of a pure substance (such as ice I, ice III, liquid water and water vapour) are not found in equilibrium at any temperature and pressure. In terms of chemical potentials there are now three equations, which cannot in general be satisfied by any values of the two variables T and p, although in principle they might be solved in a special case where one equation is mathematically dependent on the other two. In practice, however, the coexistence of more phases than allowed by the phase rule normally means that the phases are not all in true equilibrium.

Two-component Systems

For binary mixtures of two chemically independent components, $C = 2$ so that $F = 4 - P$. In addition to temperature and pressure, the other degree of freedom is the composition of each phase, often expressed as mole fraction or mass fraction of one component.

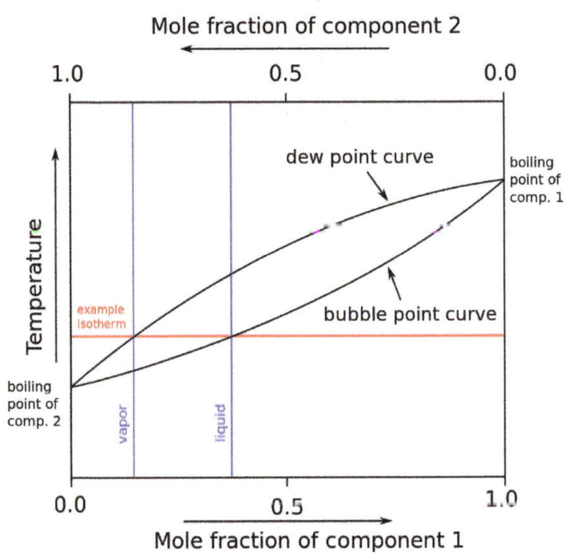

Boiling Point Diagram

As an example, consider the system of two completely miscible liquids such as toluene and benzene, in equilibrium with their vapours. This system may be described by a boiling-point diagram which shows the composition (mole fraction) of the two phases in equilibrium as functions of temperature (at a fixed pressure).

Four thermodynamic variables which may describe the system include temperature (T), pressure (p), mole fraction of component 1 (toluene) in the liquid phase (x_{1L}), and mole fraction of component 1 in the vapour phase (x_{1V}). However since two phases are in equilibrium, only two of these variables can be independent ($F = 2$). This is because the four variables are constrained by two relations: the equality of the chemical potentials of liquid toluene and toluene vapour, and the corresponding equality for benzene.

For given T and p, there will be two phases at equilibrium when the overall composition of the system (system point) lies in between the two curves. A horizontal line (isotherm or tie line) can be drawn through any such system point, and intersects the curve for each phase at its equilibrium composition. The quantity of each phase is given by the lever rule (expressed in the variable corresponding to the x-axis, here mole fraction).

For the analysis of fractional distillation, the two independent variables are instead considered to be liquid-phase composition (x_{1L}) and pressure. In that case the phase rule implies that the equilibrium temperature (boiling point) and vapour-phase composition are determined.

Liquid–vapour phase diagrams for other systems may have azeotropes (maxima or minima) in the composition curves, but the application of the phase rule is unchanged. The only difference is that the compositions of the two phases are equal exactly at the azeotropic composition.

Phase Rule at Constant Pressure

For applications in materials science dealing with phase changes between different solid structures, pressure is often imagined to be constant (for example at one atmosphere), and is ignored as a degree of freedom, so the rule becomes

$$F = C - P + 1.$$

This is sometimes misleadingly called the "condensed phase rule", but it is not applicable to condensed systems which are subject to high pressures (for example, in geology), since the effects of these pressures can be important.

References

- Çengel, Yunus A.; Boles, Michael A. (1998) [1989]. Thermodynamics – An Engineering Approach. McGraw-Hill Series in Mechanical Engineering (3rd ed.). Boston, MA.: McGraw-Hill. ISBN 0-07-011927-9

- Clapeyron, M. C. (1834). “Mémoire sur la puissance motrice de la chaleur”. Journal de l’École polytechnique (in French). 23: 153–190. ark:/12148/bpt6k4336791/f157
- Salzman, William R. (2001-08-21). “Clapeyron and Clausius–Clapeyron Equations”. Chemical Thermodynamics. University of Arizona. Retrieved 2007-10-11
- Wark, Kenneth (1988) [1966]. “Generalized Thermodynamic Relationships”. Thermodynamics (5th ed.). New York, NY: McGraw-Hill, Inc. ISBN 0-07-068286-0
- Krafcik, Matthew; Sánchez Velasco, Eduardo (2014). “Beyond Clausius–Clapeyron: Determining the second derivative of a first-order phase transition line”. American Journal of Physics. 82: 301–305. Bibcode:2014AmJPh..82..301K. doi:10.1119/1.4858403
- Carl Rod Nave (2006). “PvT Surface for a Substance which Contracts Upon Freezing”. HyperPhysics. Georgia State University. Retrieved 2007-10-16

Permissions

All chapters in this book are published with permission under the Creative Commons Attribution Share Alike License or equivalent. Every chapter published in this book has been scrutinized by our experts. Their significance has been extensively debated. The topics covered herein carry significant information for a comprehensive understanding. They may even be implemented as practical applications or may be referred to as a beginning point for further studies.

We would like to thank the editorial team for lending their expertise to make the book truly unique. They have played a crucial role in the development of this book. Without their invaluable contributions this book wouldn't have been possible. They have made vital efforts to compile up to date information on the varied aspects of this subject to make this book a valuable addition to the collection of many professionals and students.

This book was conceptualized with the vision of imparting up-to-date and integrated information in this field. To ensure the same, a matchless editorial board was set up. Every individual on the board went through rigorous rounds of assessment to prove their worth. After which they invested a large part of their time researching and compiling the most relevant data for our readers.

The editorial board has been involved in producing this book since its inception. They have spent rigorous hours researching and exploring the diverse topics which have resulted in the successful publishing of this book. They have passed on their knowledge of decades through this book. To expedite this challenging task, the publisher supported the team at every step. A small team of assistant editors was also appointed to further simplify the editing procedure and attain best results for the readers.

Apart from the editorial board, the designing team has also invested a significant amount of their time in understanding the subject and creating the most relevant covers. They scrutinized every image to scout for the most suitable representation of the subject and create an appropriate cover for the book.

The publishing team has been an ardent support to the editorial, designing and production team. Their endless efforts to recruit the best for this project, has resulted in the accomplishment of this book. They are a veteran in the field of academics and their pool of knowledge is as vast as their experience in printing. Their expertise and guidance has proved useful at every step. Their uncompromising quality standards have made this book an exceptional effort. Their encouragement from time to time has been an inspiration for everyone.

The publisher and the editorial board hope that this book will prove to be a valuable piece of knowledge for students, practitioners and scholars across the globe.

Index